SpringerBriefs in Energy

SpringerBriefs in Energy presents concise summaries of cutting-edge research and practical applications in all aspects of Energy. Featuring compact volumes of 50 to 125 pages, the series covers a range of content from professional to academic. Typical topics might include:

- A snapshot of a hot or emerging topic
- A contextual literature review
- A timely report of state-of-the art analytical techniques
- An in-depth case study
- A presentation of core concepts that students must understand in order to make independent contributions.

Briefs allow authors to present their ideas and readers to absorb them with minimal time investment.

Briefs will be published as part of Springer's eBook collection, with millions of users worldwide. In addition, Briefs will be available for individual print and electronic purchase. Briefs are characterized by fast, global electronic dissemination, standard publishing contracts, easy-to-use manuscript preparation and formatting guidelines, and expedited production schedules. We aim for publication 8–12 weeks after acceptance.

Both solicited and unsolicited manuscripts are considered for publication in this series. Briefs can also arise from the scale up of a planned chapter. Instead of simply contributing to an edited volume, the author gets an authored book with the space necessary to provide more data, fundamentals and background on the subject, methodology, future outlook, etc.

SpringerBriefs in Energy contains a distinct subseries focusing on Energy Analysis and edited by Charles Hall, State University of New York. Books for this subseries will emphasize quantitative accounting of energy use and availability, including the potential and limitations of new technologies in terms of energy returned on energy invested. The second distinct subseries connected to SpringerBriefs in Energy, entitled Computational Modeling of Energy Systems, is edited by Thomas Nagel, and Haibing Shao, Helmholtz Centre for Environmental Research - UFZ, Leipzig, Germany. This sub-series publishes titles focusing on the role that computer-aided engineering (CAE) plays in advancing various engineering sectors, particularly in the context of transforming energy systems towards renewable sources, decentralized landscapes, and smart grids.

All Springer brief titles should undergo standard single-blind peer-review to ensure high scientific quality by at least two experts in the field.

Thomas L. Brewer

Climate Action Leaders

Solutions and Opportunities

Springer

Thomas L. Brewer
Emeritus Faculty
Georgetown University
Washington, DC, USA

ISSN 2191-5520 ISSN 2191-5539 (electronic)
SpringerBriefs in Energy
ISBN 978-3-031-95666-9 ISBN 978-3-031-95667-6 (eBook)
https://doi.org/10.1007/978-3-031-95667-6

Author's Disclaimers: I have produced the book as an independent scholar. I have no financial conflicts of interest. Nothing in the book should be interpreted as offering legal or financial advice or endorsement of any products. I have used Artificial Intelligence to contribute to my reference lists, but I have not included any substantive content from AI generated materials.

This Springer imprint is published by the registered company Springer Nature Switzerland AG
The registered company address is: Gewerbestrasse 11, 6330 Cham, Switzerland

For all the Climate Action Leaders—Past, Present, and Future

Preface

The urgent need for actions to reduce greenhouse gas and particulate matter emissions has become even more obvious in the early 2020s, as extreme weather events have become more frequent and more severe, and as their consequences for human lives and economies have become more extensive. Indeed, during the time this book was being prepared, many new temperature records and related records of their effects were occurring. These recent changes in weather patterns and the climate science analyses of past and prospective developments have prompted changes in government policies, industry technologies, and economic sectors. As a result, there has been a significant increase in both the pressures and the opportunities for climate actions.

The purpose of the Policy Brief is to *stimulate increased interest* in the wide variety of solutions that have already been adapted in many countries around the world, *provide information* about the implementation and consequences of specific cases, and *facilitate international learning processes* that draw upon the experiences.

The analysis applies an interdisciplinary approach that focuses on technologies, economics, and policies, as well as the results of climate science research. The approach draws upon the work of many others—including teams of specialists in universities, think tanks, governmental agencies at all levels, and many international organizations.

The analysis includes many economic sectors—with an emphasis on energy and transportation because of their importance as sources of fossil fuel emissions. However, there are examples of actions taken in the standard IPCC lists of sectors. Among the sectors, there is a mix of geographic and governmental levels of analysis from local to global.

I have identified actions that are contributions to evolutionary processes that can reduce emissions that cause climate change. The intention is to highlight developments that have already demonstrably made significant contributions to on-going efforts to reduce the impacts of climate change through mitigation actions or could plausibly do so. The objective has not been to identify single actions that have been unquestionably the first in the world, regardless of location or sector. However, there are many examples that were the first in their sector or a specific industry within a sector. It should also be noted that the ultimate contributions of many technological contributions are still unknown as of this writing because they are part of on-going innovative processes. Alternatives to fossil fuels in the energy sector and the transportation sector are obvious examples.

There are many excellent quantitative rating systems that compare countries and aspects of their national government policies. They include much information about national governments' targets and pledges, especially including those in their Nationally Determined Contributions (NDCs) to the Paris Agreements. They also focus attention on projections of gaps between (1) what is has been pledged and what their actual measures taken are achieving, and/or (2) what has been achieved and what is needed to avoid more extreme catastrophes. These are all essential to understanding many of the problems of climate change and progress—or lack thereof—on solutions to the problems. However, they need to be complemented by information about sub-national governmental policies, especially in countries with "federalist" political systems with hundreds of regional ("state" and "provincial") governments as well as thousands of local governments around the world. Many of them have undertaken actions that not only exceed their own national governments' actions, but establish the sub-national governments as world leaders.

These features of the analysis may thus counter a defeatist reaction to proposals for action on climate change—because of a belief that the problems are so enormous they cannot be solved. The present report is based, instead, on a less defeatist approach, which is based on factual evidence about actions of many types undertaken by governments in fundamentally different socio-economic-technological circumstances as well as a variety of political systems. Please note, however, this report is not based on a "Pollyanna" approach. On the contrary, it is based on a realistic recognition that the world is already experiencing severe climate changes that have already caused catastrophic human suffering and economic losses and that the catastrophic consequences will become increasingly frequent, severe, and widespread in all regions of the world, unless there are widespread adoptions with the next few years of the diverse kinds of actions illustrated in this report.

Washington, USA Thomas L. Brewer

Acknowledgements

I am indebted to experts at Springer and many others who have made contributions throughout the editorial and production stages, especially Anthony Doyle, Acquisition Editor, and Padma Subbaiyan, Production Editor, and Lokeshwaran Manickavasagam, Project Management. Both were patient and enormously helpful from beginning to end.

I was privileged to spend the academic year 1999–2000 in Denmark at the Copenhagen Business School, where I learned a lot about Danish perspectives on climate change from faculty members, students, and business executives. In fact, a Danish business executive in a conversation with me in *1999*, said that "we understand the problem of climate change, we just have to work on the solutions." A quarter of a century later, I still marvel at such prescience—and I wonder why so many people in some countries still do not understand.

In other years, I learned a great deal on the occasions while I was a summer visitor at universities in other parts of Europe: Nuremberg University in Germany; Strasbourg University in France; and Oxford University in the UK.

Contents

List of Tables

List of Boxes

Chapter 1
Responding to the Crises: Problems, Solutions, and Opportunities

Keywords Policies · Actions · Solutions · Opportunities · Mitigation · Emissions · Fossil · Monitoring · Tracking · Tipping

1.1 Actions Needed Now

There is mounting evidence from empirical studies of the past and present, as well as projections of the future, that all regions of the world now face increasingly catastrophic consequences of climate change. Actions are therefore urgently needed for mitigation measures to reduce emissions that cause climate change. The necessary changes include a wide range of technologies and economic changes, as well as changes in policies at all levels of governance.

Excerpts from a Synthesis Report by the Intergovernmental Panel on Climate Change (IPCC 2023) document the urgency of the situation:

> Human-caused climate change is already affecting many weather and climate extremes in every region across the globe.
> Approximately 3.3–3.6 billion people live in contexts [that are] highly vulnerable to climate change.
> Some future changes are unavoidable and/or irreversible but can be limited by deep, rapid and sustained global greenhouse gas emissions reduction.
> Deep, rapid, and sustained reductions in greenhouse gas emissions would lead to a discernible slowdown in global warming within around two decades….
> Limiting human-caused global warming requires net zero CO_2 emissions.
> Maladaptation can be avoided by flexible, multi-sectoral, inclusive, long-term planning and implementation of adaptation actions, with co-benefits to many sectors and systems.

T. L. Brewer, *Climate Action Leaders*, SpringerBriefs in Energy,
https://doi.org/10.1007/978-3-031-95667-6_1

> Current global financial flows for adaptation are insufficient for, and constrain implementation of, adaptation options, especially in developing countries.
> All global modelled pathways that limit warming to 1.5 °C … with no or limited overshoot, and those that limit warming to 2 °C …, involve rapid and deep and, in most cases, immediate greenhouse gas emissions reductions in all sectors this decade.

Reminders of the urgent need for more actions are not limited to official international reports. An annual, high-level gathering of business executives and non-governmental organizations, as well as leaders from governments, have focused on the numeric discrepancy between what is needed compared with the trend in *annual changes* (World Economic Forum 2024; italics added): "The world needs immediate and systemic change. To limit global warming to 1.5 °C, annual emissions *need a 7% reduction against the current trend of 1.5% annual increase*."

The Climate Crisis Advisory Group (2023) of experts from many countries stated directly the enormity of the dangers if we remain on the paths we are following: "The world's attempts to reduce global emissions have fallen dangerously short. As a result, emissions reductions alone are no longer enough to put us back on track for a safe and manageable future for humanity."

The Global Goals organization (2024), which is dedicated to take actions to address climate change issues among the UN Sustainable Development Goals (SDGs) for 2030, has stated that "Climate change is a real and undeniable threat to our entire civilization. The effects are already visible and will be catastrophic unless we act now."

Specific temperatures exemplify the abundant data indicating the gap between the needs and pledged actions: "The collective progress of the Paris Agreement [of 2015] … demonstrated that climate plans remain insufficient to curb emissions; meeting the current Nationally Determined Contributions (NDCs) would result in a 2% decrease from 2019 levels, far from the 43% reduction needed to meet the collective goals of the Paris Agreement" (Nascimento et al. 2024, italics added; also see IPCC 2023; Liu et al. 2024; Nascimento et al. 2022).

1.2 Crises: Temperature Records, Extreme Events, and Tipping Points

Exhortations for action are reinforced by recent research results that highlight new global temperature records and related records, attributions of specific extreme weather events to climate change in the early 2020s, and emerging "tipping points."

1.2.1 Temperature Records

Global surface temperatures have also been setting new records. In 2023 the near-surface temperature was the warmest in 174-years. It was 1.45 °C above the 1850–1900 average of six temperature data sets (WMO 2023: 3). These record temperature levels are a direct result of record concentration levels of atmospheric climate change forcing agents. The three most common greenhouse gasses—i.e. carbon dioxide, methane, and nitrous oxide—have more than doubled since their pre-industrial levels in 1750. By 2022, Carbon dioxide had increased 150%, methane 264%, and nitrous oxide 124% (WMO 2024: 2). In addition, emissions of the highly potent particulate matter black carbon continue to have local as well as global impacts.

1.2.2 Extreme Weather Events

A single day's news of extreme weather events now often records the severity, diversity, and widespread locations of such events. For instance, a day in August 2024 included the temperature, flood, and wildfire events in Box 1.1.

Box 1.1 One Day's Extreme Weather Events

Chile's capital had its first rainless July on record.

China recorded its hottest month in its recent history.

July fires in *Brazil's Amazon rainforest* were a two-decade high.

Heavy rain and floods in *Pakistan* killed scores.

Bolivia had record wildfires.

In the *United States*, more than 29,000 firefighters and support personnel were trying to control 93 wildfires covering more than 2 million acres.

In California, the largest fire burned more than 600 square miles (1,550 km^2), which is larger than the city of Los Angeles.

Source: Kimathy in Reuters (2024); excerpted and edited by the author.

1.2.3 Climate Tipping Points (CTPs)

An important innovation in climate science research has been the analysis of CTPs—that is, "Climate Tipping Points". They have been defined as "elements of the [e]arth system in which small changes can kick off reinforcing loops that 'tip' a system from one stable state into a profoundly different state. … For example, a rise in global temperatures because of fossil fuel burning, further down the line, triggers a

Table 1.1 Climate tipping point estimates for collapses of the Greenland ice sheet and West Antarctic ice sheet

Element	Tipping points' degrees above pre-industrial temperature (°C)		
	Estimate	Minimum	Maximum
Greenland ice sheet	1.5	0.8	3.0
West Antarctic ice sheet	1.5	1.0	3.0

Source Kimathy in Reuters (2024); excerpted and edited by the author

change like a rainforest becoming a dry savannah. This change is propelled by self-perpetuating feedback loops, even if what was driving the change in the system stops. The system—in this case the forest—may remain 'tipped' even if the temperature falls below the threshold again" (European Space Agency 2023a, b).

At the top of a long list of "global" and "regional" CTPs are two—ice sheets in Greenland and the Antarctic—which have estimated collapse timings at only 1.5 °C above the pre-industrial level. The lower and upper estimates for both are in Table 1.1.

1.3 Emissions

In order to understand the need for actions to mitigate emissions, it is essential to focus on the types and sources of emissions—according to countries and sectors. The three biggest national emitters (China, India, and the U.S.) collectively emit slightly more than half the global total, with China alone contributing about one-third.

Of course, because the populations of China and India are about 1.5 billion each, their *per capita* emissions levels are much lower than those of the U.S., which has the highest per capita emissions at about 13 tonnes per year based on production data. The U.S. is also among the world's four highest countries in emissions per capita based on *consumption* levels.

1.3.1 Comparisons of Production-Based and Consumption-Based Emissions

There is increasing interest internationally in the differences between such production-based and consumption-based measures. Consumption-based (trade-adjusted) emissions measure domestic energy use *minus* energy used to produce exported goods, *plus* energy used to produce imported goods. The distinction between production and consumption is significant in climate change issues because of ethical questions about who is responsible for the problems.

The data in Tables 1.2 and 1.3 are the conventionally used production-based measurements. Table 1.4 compares the production and consumption emission levels for the three biggest emitters.

The inter-country differences are not trivial. For example, comparing Chinese and U.S. emissions using the commonly-used production-based emissions indicates that Chinese emissions in 2020 were 13,999 TWh or 49% greater than those of the U.S. However, consumption-based emissions for the same year were only 8261 TWh or 20% greater for China than the U.S. emissions. When developing and assessing

Table 1.2 Total GHG emissions: high-ranking countries

Countries	Total CO_2e emissions (gt)[a] in 2023	Percent of world total (gt/CO_2e) in 2023 (%)
China	12	33
US	5	13
India	2	6
EU-27	3	7

[a]CO_2e includes carbon dioxide (CO_2), methane (CH_4), and nitrous oxide (NO_2)
Source Based on excerpts by the author from Our World in Data (2025)

Table 1.3 Comparisons of major emitters' consumption and production emissions

Basis	Terawatt hours (2020)		
	China	U.S	India
Consumption	40,469	32,208	14,535
Production	42,690	28,691	14,715
Difference: Consumption–production	−2,221	+3517	−180
Ratio: Consumption/production	0.95	1.12	0.99

Source Based on excerpts by the author from Our World in Data (2025)

Table 1.4 Fossil fuel emissions and production levels of high-ranking countries

Countries	Percent of World (2023)			
	Emissions[a]	Production		
		Coal	Oil	Gas
China	32	52	5	6
United States	13	7	18	26
India	8	9	<2	<2
Russia	5	<5	12	14

[a] Includes emissions from coal, oil and gas.
Source Our World in Data (2025)

the effectiveness of policy actions, there is therefore a difference between reducing consumption-based and production-based emissions.

1.3.2 Fossil Fuel Emissions

Fossil fuel production and consumption receive special attention because there is widespread agreement among climate change experts that there must be major reductions in fossil fuel contributions to emissions over the next few years in order to avoid the worst catastrophes. Table 1.4 indicates the fossil fuel emission and production and levels of high-ranking countries.

However, there are some countries that rank high on production or consumption but not both. Most of the EU-27 countries, for instance, rank high on consumption but not production. In other regions, Iran and South Africa rank high on production but not consumption.

1.4 Economic Tipping Points

Tipping points in economic developments with positive outcomes for mitigating climate change emissions have been identified by experts at The Grantham Research Institute on Climate Change and the Environment at the London School of Economics and Political Science (2024). These TPs occur when self-reinforcing feedback processes cause further development of technological-economic changes that are beneficial to efforts to reduce greenhouse gas emissions and/or adapt to them.

The significance of such tipping points in the energy sector can be summarized from a report by Stern and Romani (2023) as follows:

> The deployment of low-carbon technologies is accelerating, with key technologies expected to reach tipping points before 2030—which in turn will trigger their scaling-up to mass market. A significant tipping point has already occurred in the energy sector, when the levelized cost of energy generation (LCOE) for solar and wind power fell below that of new coal and gas in 2018. A second tipping point, whereby the LCOE includes the cost of short-term battery storage, [was] expected to be reached in the United States in 2023, quickly followed by other countries. This rapid cost reduction led solar and wind to account for more than 75% of energy capacity additions globally in 2021.
>
> Key issues are whether, how soon, and how effectively other economic tipping points in other sectors can be achieved. Sectors that are candidates for such TPs include: transportation (trucking, aviation, and shipping, in addition to light vehicles), buildings, food and agriculture, land use, steel, cement, and chemicals (excerpted from Stern and Romani 2023).

1.5 Co-benefits for Human Health

There are numerous opportunities for actions that not only address climate change issues but also have beneficial consequences for human health. These synergistic possibilities are now at the frontiers of the climate action agenda, at the local level as well as national and international levels. Many climate actions include opportunities for cost-effective contributions to economic growth and human health.

1.6 Implications

It is increasingly clear that more substantial actions are urgently needed to avoid the most catastrophic consequences of climate change. This book analyses the past accomplishments and current agendas of climate action leaders to enhance greater efforts.

The analysis focuses on governmental policies at all levels. However, key determinants of progress include technologies and economics. Moreover, ethical issues are also increasingly important, and they interact with technological and economic determinants. The technological issues and economic issues vary across and within sectors—and so do the policy variables. In addition, fundamental industry-specific variables are obviously important, such as their ages, competitive structures, and internationally distributed production and consumption profiles. Similar contextual factors are also important in the variations among countries. The analyses throughout the book are therefore multi-dimensional.

Annex 1.1: Analytic Advancements

Although "analytic advancements" may not normally be considered "actions," they do directly increase understanding of climate change problems, solutions, and opportunities; indeed, that has been happening from the outpouring of the results of climate science research in recent years. The application of new analytic methods and the expansions of analytic domains have enriched the many aspects of climate science and provided more insights and confidence in their results. They thus provide analytic underpinnings for much of this Policy Brief.

This annex focuses on two such advancements: Measuring and Monitoring Climate Change Policies and Artificial Intelligence (AI). There is a central difference in their analytic tool kits: Whereas the former relies principally on established statistical methods, the latter is driven by different approaches to underlying data processing. Despite differences in their analytic methods, they both provide additional information about climate change problems, solutions, and opportunities, as well as climate action leaders.

A.1 Measuring and Monitoring Climate Change Policies

There are easily accessed international-comparative data collections and related analyses for tracking policies over time. They include several developed by major international organizations; there are also others developed by climate change programs associated with think tanks and universities and yet others by international news organizations. They are summarized here and applied in subsequent chapters.

The Climate Actions and Policies Measurement Framework (CAPMF)

The Climate Actions and Policies Measurement Framework (CAPMF) has been developed by the OECD in collaboration with the IEA, ITF, and NEA (Nachtigall et al. 2022). The framework includes 128 policy variables, which are used to analyse the patterns and trends of policy adoptions of 52 countries over the period 2000–2020 and beyond for each year. The framework includes sectoral, cross-sectoral and international actions and policies. To date, the focus has been on mitigation policies, with an on-going process to include more adaptation policies as well. The data include the "adoptions" and "stringencies" of enacted laws. The data to date do not include all types of policies, nor measures of "implementation or enforcement" or "effectiveness." However, the broad conceptual framework, with the available data, can be applied to countries' policies for both empirical and evaluative analyses.

The data set includes 38 OECD members plus 14 "partner countries" that include OECD membership accession candidates, G20 countries not in the OECD, and the European Union as a collective unit. The 52 countries collectively emit more than 80% of global ghg emissions, and each of the 52 countries individually emits more than 1% of the global total.

Climate Action Tracker (CAT)

Climate Action Tracker focuses on the effects of national emissions policies. It covers about 85% of world emissions and 70% of world population. It includes examples of specific tangible features of many policies.

Globe

Globe is a multi-regional dataset that aims to monitor energy trends, emissions transitions, and building stocks on a global scale. The GLOBE climate legislation study reviews climate change legislation enacted in 66 countries (Nachmany et al. 2014).

The Climate Change Performance Index (CCPI), CCPI

The Climate Change Performance Index (CCPI), published annually since 2005, is an independent monitoring tool for tracking countries' climate mitigation performance. It increases transparency in national and international climate policy and enables comparison of individual countries' climate action efforts and progress. The CCPI indicates the climate mitigation efforts of 63 countries and the EU, which collectively account for over 90% of global greenhouse gas (GHG) emissions. Results are compiled with the aid of 450 national experts evaluating their countries' most recent national and international climate policies.

A.2 Artificial Intelligence (AI)

Applications of Artificial Intelligence (AI) to climate change have become salient and contentious topics in many contexts. In the present context, key issues pertain to: (1) the scope and significance of climate change topics where AI can be or should be applied, and (2) the increased emissions of greenhouse gases and particulate matter caused by AI's increasing electricity consumption in data processing centers.

As for AI applications to climate change topics, there have been suggestions of many kinds of actual and potential contributions. Harrison (2023) has found examples of improving the prediction of severe weather events and the measurement of soil carbon stock changes in agriculture and forestry practices.

Anderson (2024) has identified the following possibilities: improved analysis of weather data and electricity demand patterns that can "optimize the operation of renewable energy sources like wind and power;" improve monitoring of deforestation; develop more energy efficient building designs and more low-emission transportation routes.

Another set of possibilities extracted from Rolnick et al. (2019) and reported by Hao (2019) includes AI applications that improve: predictions of electricity demand; discovery of new energy-saving materials; optimization of how freight is routed; electric-vehicle adoption rates; buildings' efficiency; supply chain optimization; energy consumption estimates; the proportions different types agricultural of crops; tracking deforestation; consumers' shopping behavior.

Minevich (2021) identified specific projects in seven countries by consulting organizations that are applying AI applications to a wide range of sectors. Although the focus of the projects is the measurement of businesses' emissions, there are issues that are common for both governments and businesses, as exemplified in the following observations from a representative of the Boston Consulting Group: "Corporations' climate action does not match our ambitions: one of the main reasons is that corporations don't accurately measure their climate impact and track it enough over time. Fortunately, tools and techniques to help us exist[;] one of the most powerful ones is artificial intelligence." Furthermore, "we must reduce these emissions by 50% by the end of this decade."

Concerns about the increases in emissions from AI consumption of electricity in data enters have been addressed, for example, in Goldman Sachs Research (2024). Their projection is that data centers' production of the total world electricity production will increase by about 160% by 2030, compared with 2023. During that period, AI's consumption will increase by about 200 terawatts *per year*. Such increases would make the AI share of total data center consumption about 19% in 2028. Of course, the actual future amounts will depend on the complex interplay of technological, economic, and political developments. Already, a salient case has been the re-commissioning of a shut-down Australian coal-fired power plant specifically to provide electricity for AI project consumption.

References

Anderson K (2024) How can artificial intelligence help tackle climate change? https://greenly.earth/en-us/blog. Accessed 16 Feb 2025

Armstrong McKay D et al (2022) Exceeding 1.5 °C global warming could trigger multiple climate tipping points. Science 377:6611. https://doi.org/10.1126/science.abn7950. Accessed 5 May 2024

Climate Crisis Advisory Group (2023) Climate interventions: exploring sustainable and innovative pathways for humanity. https://www.ccag.earth/news/ccag-calls-for-urgent-scientific-examination-of-vital-climate-interventions. Accessed 1 June 2024

European Space Agency (2023a) Understanding climate tipping points. https://www.esa.int/Applications/Observing_the_Earth/Space_for_our_climate/Understanding_climate_tipping_points. Accessed 16 Aug 2024

European Space Agency (2023b) Global stocktake: how space drives climate action. https://climatedetectives.esa.int/climate-change-kit-2/. Accessed 16 Aug 2024

Global Goals (2024) Climate action. https://www.globalgoals.org/13-climateaction. Accessed 8 Aug 2024

Goldman Sachs Research (2024) AI is poised to drive 160% increase in data center power demand. https://www.goldmansachs.com/insights/articles/AI-poised-to-drive-160-increase-in-power-demand. Accessed 8 Mar 2025

Grantham Research Institute on Climate Change and the Environment at the London School of Economics and Political Science (2024) Our vision. https://www.lse.ac.uk/granthaminstitute/. Accessed 1 June 2024

Hao K (2019) Here are 10 ways AI could help fight climate change. https://www.technologyreview.com/2019/06/20/134864/ai/climate-change-machine-learning/. Accessed 20 June 2019

Harrison, K. (2023) AI revolutionizes climate science: predicting storms, climate trends, and sustainable land use. https://www.environmentenergyleader.com/stories/ai-revolutionizes-climate-science-predicting-storms-climate-trends-and-sustainable-land-use,4621? Accessed 8 Mar 2025

Intergovernmental Panel on Climate Change (IPCC) (2023) Synthesis report of the IPCC Sixth Assessment Report (AR6). Used with permission. https://report.ipcc.ch/ar6syr/pdf/IPCC_AR6_SYR_SPM.pdf. Accessed 25 May 2024

Kimathy S (2024) Sustainable switch, climate focus. Reuters, August 2. Excerpts used with permission. https://www.reuters.com/. Accessed 5 Aug 2024

Liu Z, Deng Z, Davis SJ, Ciais P (2024) Global carbon emissions in 2023. Nat Rev Earth Environ 2024. https://doi.org/10.1038/s43017-024-00532-2AccessedMay5

Minevich M (2021) 11 examples of AI climate change solutions for zero carbon. https://www.forbes.com/sites/mrkminevich/2021/10/08/11/. Accessed 8 June 2024

Nascimento L, Godinho C, Kuramochi T, Moisio M (2024) Michel den Elzen & Niklas Höhne (2024) Climate policy in 2023. Nat Rev Earth Environ 5:255–257. https://doi.org/10.1038/s43017-024-00541-1AccessedMay5

Nascimento L, Kuramochi T, Höhne N (2022) The G20 emission projections to 2030 improved since the Paris Agreement, but only slightly. Mitig Adapt Strateg Glob Chang 27:39. Accessed 5 May 2024

OECD (2023) OECD AI principles overview. https://oecd.ai/en/ai-principles. Accessed 7 May 2023

Our World in Data (2025) Data Catalog. https://ourworldindata.org/data?topics=Energy+and+Environment~CO2+%26+Greenhouse+Gas+Emissions. Accessed 3 May 2025

Rolnick D, Donti P et al (2019) Tackling climate change with machine learning. Comput Sci. https://arxiv.org/abs/1906.05433. Accessed 10 Apr 2023

Stern N, Romani M (2023) The global growth story of the 21st century: driven by investment and innovation in green technologies and artificial intelligence. https://www.lse.ac.uk/granthaminstitute/wp-content/uploads/2023/01/The-global-growth-story-of-the-21st-century-driven-by-investment-in-green-technologies-and-AI.pdf. Accessed 1 June 2024

World Economic Forum (2024) the state of climate action: major course correction needed from + 1.5% to −7% annual emissions. https://www.weforum.org/publications/the-state-of-climate-action-major-course-correction-needed-from-1-5-to-7-annual-emissions/. Accessed 5 May 2024

World Meterological Organization (2023) https://wmo.int/publication-series/state-of-global-climate-2023. Accessed 5 May 2024

Chapter 2
The Nordic-5

Keywords Nordic · Denmark · Finland · Iceland · Norway · Sweden · Wind · Partnerships · EVs · Shipping

2.1 Introduction

The Nordic Council has established themes and ambitions for the group of five member countries, and it has tracked key features of the countries' emissions and policies. It has also sponsored numerous studies on diverse topics of common interest (Nordic Council of Ministers 2023a, b, 2024a). As subsequent sections of the chapter focus on the actions of the individual countries, however, it is evident that the varying natural resources, industrial specialties, and other features of the countries are important determinants of the distinctive emphases in their policies. As a Nordic Council of Ministers report (2023a, b; also see 2024b, d, e) noted: "Reaching climate neutrality is a massive task, but there are multiple pathways to climate neutrality in the Nordic countries."

Nevertheless, there is a recurrent theme in the Nordic Council and individual member countries' programs: they work to be world leaders in climate action. As the Secretary-General of the Council said at a COP meeting (Nordic Council 2024c): "The Nordic Co-operation is aiming not only at transforming the Nordic region but also providing a blueprint for the world."

The Nordic countries have in fact made significant reductions in greenhouse gas emissions, as indicated by the data in Table 2.1. Based on per capita CO_2-equivalent emissions, the percentage decline for each of the Nordic countries over the 1990–2022 period was much greater than the global average. Denmark exhibited the largest decline at 45%, compared with the global average of 4%.

T. L. Brewer, *Climate Action Leaders*, SpringerBriefs in Energy,
https://doi.org/10.1007/978-3-031-95667-6_2

Table 2.1 Changes in nordic countries' GHG emissions (per capita CO_2e)

Nordic country[a]	Per Capita CO_2e gt[c]		Change (1990–2022)	
	1990	2022	CO2e gt	Percent
Denmark[b]	13.7	7.6	−6.1	−45
Finland[b]	18.1	13.1	−5.0	−28
Iceland	14.2	13.0	−1.2	−8
Norway	14.4	13.4	−1.0	−7
Sweden[b]	8.0	6.7	−1.3	−16
Global	7.1	6.8	−0.3	−4

[a]All five are members of the Nordic Council
[b]Denmark, Finland, and Sweden are members of the EU-27; Iceland and Norway are not EU members. However, all five countries participate in the EU Emission Trading System (ETS)
[c]Includes land use change
Source Derived from Ritchie et al. (2020, rev. 2024)

The data in Table 2.1 are the conventionally used production-based measurements. There is increasing interest internationally in the differences between production-based and consumption-based measures. Such comparisons are made in Table 2.2 for four of the Nordic countries.

Finland is notable for its relatively low consumption level compared with its production level, thus indicating that its exports of energy exceed its imports. Denmark, on the other hand, reveals the opposite. The origins of the differing patterns appear in the individual country sections below.

Table 2.2 Comparisons of consumption and production emissions

Country	GHG emissions: energy—TWh (2020)			
	Consumption	Production	Consumption—production	Consumption/production
Denmark	337	274	+63	1.23
Finland	441	472	−31	0.93
Norway	424	412	+12	1.03
Sweden	579	577	+2	1.00

Consumption-based (trade-adjusted) energy use measures domestic energy use minus energy used to produce exported goods, plus energy used to produce imported goods
Iceland data not in source
Source Excerpted and computed by the author from data in Ritchie et al. (2020, rev. 2024)

2.2 Denmark

A ranking of 63 countries by the Climate Change Performance Index (2024) listed Denmark number 1 in the world. Its summary score of 78.37 was 69.60 points above second place Netherlands. A detailed report by the International Energy Agency (2024a) has summarized Denmark's accomplishments as a climate action leader, as in Box 2.1. Also see Danish Government (nd, 2021, 2024).

Box 2.1 Denmark's Climate Action Record

Denmark has been an early leader in decarbonisation .... Denmark's technology leadership is important in the areas of offshore wind, biomethane and district heating. The government has expanded these categories to adopt a strategic focus on carbon capture and storage (CCUS) and hydrogen. The government has a robust energy and climate governance under the Danish Ministry of Climate, Energy and Utilities and ... the Climate Act of 2020 ensures annual policy actions and funding. Among IEA countries, Denmark has the highest share of wind electricity, which together with bioenergy and solar photovoltaic make up over 80% of the electricity mix. The district heating sector has practically phased out coal, contributing to lower reliance on fossil fuels in Denmark's total energy supply than the IEA average.

Source: IEA (2024a)

Sorensen and Clausen (2023) have put the Danish record in a technological and political context: "The combination of political ambition and technological know-how has made Denmark a both real and self-proclaimed climate front runner. [Moreover], a trade-off between climate action and welfare is not generally recognised by the Danes. To the contrary, 68% of Danes understand the green transition rather as a driver of growth and welfare." On climate change issues, there is a public consensus that supports many actions, generally with relatively small differences among parties and age groups.

Table 2.3 provides evidence of Denmark's progress in transitioning to a more sustainable energy sector, where its widely recognized contribution of wind is apparent. During the period from 2000 to 2023, fossil fuels declined from more than 80% of the total to about 10%. Over the same period, wind increased from about 10% to nearly 60%, and solar PV increased from virtually zero to 10%.

The Global Climate Legislation Study (Nachmany et al. 2014) summarized Denmark's approach to climate action as: "Sector action plans outline specific environmental objectives and measures to achieve the objectives and deadlines, which allow the evaluation of their effectiveness." An archive of hundreds of "solutions" in about 80 "focus areas" is indicative of the wide range of solutions as well as problems that can be considered (Danish Government, State of Green nd; Sorensen and Clausen 2023). There are, however, two sectors where progress has been less significant, namely transportation and agriculture. Yet, even in these sectors there

Table 2.3 Changes over time in Denmark's energy sector sources

Sources	2000		2023	
	GWh	% of total	GWh	% of total
Coal	16,673	46.2	2524	7.4
Oil	4,439	12.3	312	0.1
Natural gas	8,774	24.3	1000	2.9
Subtotal: Fossil fuels	29,866	82.8	3,836	11.3
Wind	4,241	11.8	19,539	57.6
Solar PV	1	<0.1	3,363	9.9
Subtotal: Wind and solar	4,242	11.8	22,902	67.5
Biofuel	620	<0.1	5512	16.2
Other	39	<0.1	–0–	–0–

Source Compiled by the author from data in IEA (2023a, 2024a)

have been recent actions that make Denmark a world leader on some key issues. It has achieved a world first in maritime shipping with international ferry service using fully electric propulsion that replaces diesel-powered engines. In the agriculture sector, another world first is the imposition of a fee for the emissions of methane by cows and other livestock. In 2024, the government introduced a reforestation project that will transform 10% of the current agricultural land into forests.

An important feature of Denmark's approach to climate actions has been to develop partnerships between government and business based on a program established in 2019, "Climate Partnerships for a Greener Future" (Danish Government, State of Green 2021). During the next few years, partnerships in 14 sectors—such as energy, transportation, construction, and agriculture—developed more than 400 recommendations for achieving a 70% reduction in ghg emissions by 2030. In February, the government indicated that it was on track to achieve the 70% reduction (Rassmussen 2025). It can be noted, however, that the basis is the commonly-used "territorial" indicator of domestic production emissions, which are less than the "consumption" basis, or international travel emissions—both of which are relatively large in Denmark and many other high income countries. The cases in Box 2.2 are evidence of how specific actions resulted in reduced ghg emissions in the first few years.

Box 2.2 Action Cases in Denmark's Climate Partnership Program

- Maritime shipping company introduced a container vessel fueled by carbon neutral methanol, with eight more scheduled to follow.

- Preparation of the land for construction of a sustainable district was done using only fossil-free processes.
- Utilities in Copenhagen use a carbon capture and storage system for 3 million tonnes of CO_2 annually.
- A cement factory has a heat recovery system with a heat exchanger that transfers thermal energy from the flue gas to a local city district heating network.
- Nordic Development Financing Institutions provided equity and private and development banks provided loan financing for EUR 600 million for a 310 MW wind farm in Africa.
- Biomass replaced crude oil in the manufacture of plastics to establish compostable material.
- New water valve technology replaced an old system to reduce emissions in a brewery.
- A data energy center redistributed surplus heat to a local city's district heating system.

Source: Danish Government, State of Green (2021).

The Danish partnership program was internationalized in 2024 at COP29 by an agreement with the World Business Council for Sustainable Development (WBCSD). The international partnership is designed to use Denmark's experience with business-government partnerships in the WBCSD's network of more than 230 businesses in countries around the world (Danish Government, State of Green 2024).

In short, Denmark has established a record as a climate action leader in the energy sector and to a lesser extent in other sectors as well. It is not only a leader among European countries but all countries. In fact, its potential for serving as an example for international learning processes is manifest in a report specifically about *Electrification of U.S. Industry: Applying Lessons from Denmark* (Hoffmeister et al. 2024). They concluded that "The United States can accelerate into a new era of industrial resilience by applying lessons learned from our allies, such as Denmark. We can build on existing collaboration between stakeholders, including policymakers, utilities, industrial firms, technology vendors, and others." Among the many lessons were:

- Strong, clear national policy (and international standards and collaboration) drives industrial electrification in Denmark.
- A centralized, independent testing facility supports industry through the energy transition. Companies rely on their results for performance and financial returns.
- Dynamic electricity pricing, a national energy data hub, and regional interconnections have been other important drivers of Danish electrification.
- Public–private partnerships between manufacturers and government are critical for accountability, government understanding of industry needs, and engagement.

A Danish business executive has explained the disproportionate worldwide success of Danish companies in many economic sectors, relative to non-Nordic firms in the rest of the world: "It's not necessarily the strongest that survives but the one that will adapt to changes" (Economist 2024). It could be an aphorism for governments as well as businesses, as they decide how to respond to climate change issues.

In any case, Denmark is not alone among the Nordic countries as a world leader, where governments as well as businesses have adapted to climate change by changing their own emission reduction policies, as the following sections about other Nordic countries demonstrate.

2.3 Finland

Among the five Nordic countries, Finland was second for its reduction of per capita CO2e emissions by 28% compared with a global average of 4% over the period from 1990 to 2022.

A significant contributing factor to its low emission rate is its relatively low usage rate of fossil fuels. For instance, in 2021 oil was only 21% of total energy production, and all fossil fuels together were 35%. The bulk of the production came from less carbon intensive sources: hydro, wind, and nuclear. Thanks to its nuclear reactors and large domestic production of renewable energy (mainly forestry solid biomass as well as generation from hydro and wind), Finland has one of the lowest levels of reliance on fossil fuels among IEA member countries (IEA 2023b). It is also a world leader in key technologies—especially batteries and nuclear reactors.

Finland is perhaps best known for the progression of provision in its Climate Change Acts beginning in 2015—and updated in 2022—with a *legal obligation* to reach carbon neutrality by 2035. This commitment makes it a world leader for the short-term net zero target. Recent changes in government administration's, however, have inhibited a long-term consistency in its policy actions.

2.4 Iceland

Iceland is unusual among the Nordic countries in the share of its emissions from the Land Use/Land Use Change/Foresty (LULUCF) sector. In fact, LULUCF emissions have accounted for about two-thirds of its annual total emissions since 1990, and as recently as 2023 its total emissions had *increased* about 6 from 1990 to 2023 (Nordic Council of Ministers 2023a, b).

The sectoral sources of the emissions are summarized in Table 2.4. A substantial portion of the relatively large contribution of industry can be attributed to growth of the aluminum industry. Road transport emissions also increased, but that trend is in

Table 2.4 Sectoral sources of Iceland's emissions

Sector	Percent of emissions (kt CO_2eq. 2021)
Industry	43
Heat and electricity	4
Road transport	19
Agriculture	13
Fisheries	12
Other[a]	9

[a] Added by the author
Source Nordic Council of Ministers (2023a, b)

the process of being reversed by the increasing use of electric vehicles. The shares of emissions from agriculture and fishing have been decreasing.

As for mitigation policies, Iceland is well known for its system of geothermal systems that heat Reykjavik and other areas of the country (Tómas 2024; Mims 2008). The first geothermal system may actually extend back nearly a thousand years, according to the Icelandic Sagas. In any case, today's systems are used to heat 90% of the homes. The sources of the heating systems are in the combination of underground water reservoirs and hot magma that heats the water to as much as 400 °C. In its totality, Iceland has the largest district heating system in the world. Because the geothermal heat is also used to produce electricity, in combination with its extensive hydroelectric energy sources, 99% of Iceland's electricity is produced by renewable sources. As the world leader in geothermal technology, Iceland has transferred the technology to cities in China, Germany, and the United States.

In the transport sector, Iceland has taken several policy actions to support the purchase and operation of electric vehicles (EVs). They include reductions in the Value Added Tax (VAT) for the purchase of plug-in hybrid and electric cars; free parking for EVs in the city centre of the capital Reykjavik and also in the city center of Akureyri; and subsidies for EV charging stations (IEA 2024b).

2.5 Norway

Norway's climate change profile includes an unusual combination of features. One feature is the low level of ghg emissions from its energy sector because of the country's extensive hydropower facilities, which produce about 90% of its electricity. Another feature is the production of fossil fuels; Norway is a major producer of oil and natural gas—most of it for export. It also is developing several large offshore wind farms—one of them among the largest in the world. In the transportation sector, it is a world leader in the adoption of electric vehicles.

As indicated in Table 2.1, Norway's per capita ghg emissions have been higher than Denmark's and Sweden's but about the same as Finland's and Iceland's; they declined by 7% from 1990 to the early 2020s. The Climate Change Act of 2017 established emission targets, including a 90–95% reduction by 2050 compared with 1990. It already had in place in 1991 one of the world's first carbon taxes on fossil fuels and the petroleum industry.

During the three decades from 1990 to the early 2020s, Norway's GDP increased by about two-thirds during the same period that its emissions declined by one-third. It is thus among the world leaders in decoupling emission reductions and GDP growth.

Norway is emerging as a world leader in two modes of transportation: electric automobiles and electric airplanes. By the end of 2024, Norway had the highest percentage in the world of EVs in new automotive purchases (Murray 2025). New policies began in the early 1990s, and they have included taxes on gasoline and diesel vehicles, but not on EVs and other measures such as higher taxes and registration fees for fossil fuel vehicles. It has also made much progress in its plan to use only battery-powered aircraft by 2040 on its short-haul flights. The government corporation that operates 45 airports has been ordering such planes because they can accelerate faster on short runways than traditional planes, and they can easily recharge for short flights over fiords. Many other actions are summarized in Box 2.3.

Box 2.3 Norway's Climate Action Record

Although not a member of the EU, Norway participates in the EU Emissions Trading System (ETS), its Effort Sharing Regulation (ESR) for non-ETS emissions, and the land-use, land-use change, and forestry regulation system (LULUCF).

The Norwegian Climate Action Plan for 2021–2030 is reducing emissions in the transport, buildings, waste, and agriculture sectors by 45% in 2030 compared with 2005.

The Plan also raised the carbon tax rate to the equivalent of 196 Euros per tonne of CO_2-equivant emissions in 2030.

Beginning in 2010, building energy efficiency certificates for buildings when they are built, rented, or sold.

Since 2016, the installation of fossil fuel heating systems has been prohibited.

Fossil fuel automobiles are subject to a special registration tax, CO_2 tax, and road use tax on diesel and gasoline fuel vehicles.

Electric vehicles (EVs) are supported by not being subject to a registration tax or a value added tax (VAT), and by qualifying for reduced toll road fees, ferry fees, and parking fees. [In 2023, 100% of the new autos purchased in Norway were electric vehicles.]

Source: Extracted, edited, and amplified by the author from IEA (2024c).

Norway is also a leader in several frontier technologies, one of which is Carbon Capture and Storage (CCS), for which it has several projects in progress. It anticipates exporting CCS services. Another technology where it has projects in progress is low-carbon hydrogen. Such capabilities and interests have led the IEA (2024c) to recommend to the Norwegian government that it should: "Increase ambitions to jump-start clean technologies where Norway may have competitive advantages and means, such as hydrogen, green shipping, carbon capture and storage, and offshore wind."

2.6 Sweden

Sweden's total CO_2 emissions declined by 38.2% from 1990 to 2023, and its per capita emissions in 2023 were the third lowest among all IEA countries. A key event during this period was the establishment in 2015 of a platform named Fossil Free Sweden (FFS). The FFS platform includes roadmaps for 22 sectors that are designated for cooperative decarbonization actions by businesses and local governments as well as the national government. The coverage is broad; the sectors were the source of 70% of the country's ghg emissions in 2022. Box 2.4 highlights developments in the electricity, transportation, industry, and buildings sectors.

Box 2.4 Sectoral Examples of the Fossil Free Sweden Platform

Electricity. The electricity sector has been fueled by hydro (40%), nuclear (30%), and wind (about 20%) in the early 2020s. Fossil fuels contributed less than 1% in 2022.

Transport. It is essential to distinguish among road, aviation, and maritime transport sub-sectors. As for road transportation, in 2018 Sweden established subsidies to purchase new low-carbon passenger cars, light buses and light trucks; it also established a tax for gasoline and diesel vehicles. Although the low-carbon vehicle subsidies were phased out in 2023, the low tax rates for low-carbon vehicles have been maintained. A temporary car scrappage scheme to replace older cars with EVs in 2024–25 was introduced. Such actions had intended results: EV sales increased from 8% in 2018 to about 60% in 2023. Meanwhile, sales of gasoline and diesel vehicles declined from 85 to 31%. Local measures such as congestion charges and traffic regulations in major cities and public transport subsidies aim to encourage other forms of transport over driving. For instance, in Gothenburg a 'Green Zone' is being developed in the city.

Aviation. There are special taxes for the aviation industry, including not only domestic flights but also international flights departing from or arriving in Sweden. There are also biofuel blending requirements.

Maritime Shipping. Sweden has been a leader in monitoring ships' emissions—in cooperation with Danish monitoring organizations—as they sail between the Baltic Sea and the North Sea. It has also been a leader on climate change issues in the International Maritime Organization (IMO), and it has been a leader in imposing environmental taxes for maritime shipping.

A project that encourages a "green industrialization transformation" in the far northern region near the Arctic circle is already creating thousands of new employment opportunities in the industry and electricity sectors (Milne 2024). Also, in particular, the national steel industry is being transformed by the use of fossil-free hydrogen.

Government building regulations, energy performance certificates, and subsidies for homes that switch from fossil fuel heating—all these are reducing dependence on fossil fuels and increasing energy efficiency.

Sources: Excerpts adapted from IEA (2024d).

As for the future, a significant challenge for Sweden is to reverse the expected increase in transport emissions, including both the road and aviation subsectors. The target for transport emissions in 2030 is 70% less than in 2010. What action will achieve that target remains to be evident as of early 2025. As for aviation, it remains to be seen whether Sweden can diverge from the world projections. Changes in 2025 indicated slowing down and reversing policy directions in several aspects.

2.7 Summary Comparisons

Table 2.5 presents quantitative summary country-based rankings based on the Climate Change Performance Index (CCPI).

Table 2.5 Climate Change Performance Index (CCPI) for Four Nordic Countries

Country[a]						
	Overall numeric world rank[b]	Policy rating[c]	Renewable energy[c]	Energy use[c]	GHG emissions[c]	Overall ratings[c]
Denmark	1	High	Very high	Medium	High	High
Finland[d]	34	Low	High	Very low	Medium	Low
Norway	6	Low	Very high	Very low	Medium	High
Sweden	8	Low	Very high	Very low	High	High

[a]The total number of countries included in the CCPI is 63 + the EU, which "account for over 90% of global greenhouse gas emissions." Iceland is not included in the CCPI

[b]2025 country rankings (based on 2024 data) among 63 countries. The rankings in the table therefore reflect the rankings among the ranked countries. For example, Denmark was fourth because none of the 63 countries achieved the three potential *very high* rankings in the ranking system

[c]Potential ratings: Very High, High, Medium, Low, Very Low

[d]Finland's world rank the previous year was 11 places higher

Source Excerpts by the author from Climate Change Performance Index (CCPI) (2024)

Although three are among the top ten of the 63 countries and four are "high" or "very high" in "Renewable Energy," the Nordic countries are otherwise quite different. The difference in the "Policy" ratings is evident, with Denmark in number 1, as it was in the previous year, compared with Finland, which was down 11–34th from 23rd the previous year. Changes in the governments of Finland and Sweden had also resulted in significant changes in the approaches to climate change issues. Yet, in the broader context of the ratings of countries all over the world, three are among the top ten.

Annex 2.1: Energy Efficiency Policies in Nordic Countries' Buildings and Industry Sectors

This annex includes comparative summaries of Nordic Policies in the buildings sector and industry sector derived from a report by the Nordic Council (2025). The favorable summary statement about the advantages of the *synergistic* and *integrative* approach that has been adopted is evident in Box 2.5.

Box 2.5 Advantages of the Multi-policy Approach

The synergistic application of these instruments creates a supportive environment for the widespread adoption of energy efficiency measures. *Public policies* establish enforceable frameworks that ensure compliance through regulations, incentives, and minimum standards. However, they can be slow to adapt, potentially limiting innovation. In contrast, *voluntary initiatives* offer greater flexibility, fostering innovation and encouraging practices that exceed compliance standards. However, they may face challenges with inconsistent participation. The most effective strategy integrates both approaches—using voluntary initiatives to drive innovation and public policies to scale these innovations, ensuring broad impact and alignment with regulatory frameworks.

Source: Nordic Council (2025)

Specific examples of building sector policies are in Panel A and Industry sector policies in Panel B of Table 2.6.

Table 2.6 Nordic energy efficiency policies[a]

Policy types	Denmark	Finland	Iceland	Norway	Sweden
Panel A. Bulding Sector					
Regulatory					
Energy efficiency requirements	Yes	Yes	Yes	Yes	Yes
Energy labelling systems	Yes	Yes	–	Yes	Yes
Financial					
Grants, tax reductions for energy efficiency measures, heat pumps, window replacements	Yes	Yes	Yes	Yes	Yes
Voluntary initiatives					
Eco label certifies energy efficient building	Yes	Yes	Yes	Yes	Yes
Informational instruments					
Advisory services available	Yes	Yes	Yes	Yes	Yes
Panel B. Industry Sector					
Regulatory					
Mandatory	Yes	Yes	–	Yes	Yes
Grants, loans	Yes	Yes	Yes	Yes	Yes
Financial					
Grants, tax reductions for energy efficiency measures, heat pumps, window replacements	Yes	Yes	Yes	Yes	Yes
Voluntary Initiatives					
Eco label certifies energy efficient building	Yes	Yes	Yes	Yes	Yes
Informational Instruments					
Advisory services available	Yes	Yes	Yes	Yes	Yes

[a]Faroes Islands, which are an independent region of Denmark, are reported separately as members of the Nordic Council.

Source Excerpted by the author from Nordic Council (2025)

References

Climate Change Performance Index (CCPI) (2024). https://ccpi.org. Accessed 1 Mar 2025

Danish Government, State of Green (nd) Solutions. https://stateofgreen.com/en/solutions/. Accessed 14 Nov 2024

Danish Government, State of Green (2021) Climate partnerships for a greener future. https://stateofgreen.com/en/wp-content/uploads/2022/04/Climate-Partnerships_WEB-komprimeret.pdf. Accessed 14 Nov 2024

Danish Government, State of Green (2024) COP29: new agreement set to internationalise Denmark's climate partnerships. https://stateofgreen.com/en/solutions/. Accessed 14 Nov 2024

Economist (2024) Why are Nordic companies so successful? December 30. https://www.economist.com/business/2024/12/30/why-are-nordic-companies-so-successful. Accessed 31 Dec 2024

Hoffmeister A, Hart R, Chen H (2024) Electrification of U.S. industry: applying lessons from Denmark. American Council for an Energy-Efficient Economy (ACEEE). https://www.aceee.org/white-paper/2024/12/electrification-us-industry-applying-lessons-denmark. Accessed 9 Dec 2024

IEA (2023a) Denmark 2023, energy policy review. http://www.iea.org. Accessed 2 Nov 2024

IEA (2023b) Finland 2023, energy policy review. http://www.iea.org Accessed November 2, 2024.

IEA (2024a) Energy system of Denmark. https://www.iea.org/countries/denmark. Accessed 18 Oct 2024

IEA (2024b) Policies data base: Iceland. https://www.iea.org/policies?country=Iceland. Accessed 5 Dec 2014

IEA (2024c) Energy policy review 2022. Norway. https://iea.blob.core.windows.net/assets/de28c6a6-8240-41d9-9082-a5dd65d9f3eb/NORWAY2022.pdf. Accessed 4 Dec 2024

IEA (2024d) Sweden: 2024, energy policy review. http://www.iea.org Accessed 9 Dec 2024

Mims C (2008) One hot island: iceland's renewable geothermal power. Sci Am, October 20. https://www.scientificamerican.com/article/iceland-geothermal-power/. Accessed 5 Dec 2024

Murray A (2025) Norway on track to be first to go all-electric. https://www.bbc.com/news/articles/cg52543v6rmo. Accessed 14 Jan 2025

Nachmany M et al. (2014) The globe climate legislation study: a review of climate change legislation in 66 countries. https://www.academia.edu/7301094/The_GLOBE_climate_legislation_study_a_review_of_climate_change_legislation_in_66_countries?email_work_card=title. Accessed 16 Feb 2025

Nordic Council (2025) Energy efficiency in the Nordics. https://pub.norden.org/nordicenergyresearch2025-01/nordicenergyresearch2025-01.pdf. Accessed 19 Feb 2025

Nordic Council of Ministers (2023a) Nordic stocktake: pathways to climate neutrality. https://www.norden.org/en/publication/nordic-stocktake-pathways-climate-neutrality. Accessed 15 Nov 2024

Nordic Council of Ministers (2023b) Nordic stocktake: pathways to climate neutrality. https://norden.diva-portal.org/smash/get/diva2:1807947/FULLTEXT01.pdf. Accessed 15 Nov 2024

Nordic Council of Ministers (2024a) Sustainable living hub. https://www.norden.org/en/publication/sustainable-living-hub. Accessed 15 Nov 2024

Nordic Council of Ministers (2024b) Nordic toolbox for a just green transition. https://www.norden.org/en/publication/nordic-toolbox-just-green-transition. Accessed 15 Nov 2024

Nordic Council of Ministers (2024c) Nordic climate action and the road to COP29. https://www.norden.org/en/news/nordic-climate-action-Road-to-COP29. Accessed 15 Nov 2024

Nordic Council of Ministers (2024d) Nordic climate actions: showcasing scalable solutions aligned with the global stocktake decision objectives. https://doi.org/10.6027/nord2024-040. Accessed 15 Nov 2024

Nordic Council of Ministers (2024e) Nordic summit 2024. https://nordicsummit.info/agenda/#sz-tab-45563. Accessed 15 Nov 2024

Rasmussen L (2025) Denmark on track to hit 2030 emissions cuts goal, council says. Reuters. https://www.reuters.com/sustainability/climate-energy/denmark-track-hit-2030-emissions-cuts-goal-council-says-2025-02-27/. Accessed 27 Feb 2025

Ritchie H, Rosado P, Roser M (2020) Energy production and consumption. Published online at OurWorldinData.org. https://ourworldindata.org/energy-production-consumption. Accessed 2 Nov 2024

Sørensen D, Clausen J (2023) Denmark: climate as a given. In: Kaeding M et al (eds) Climate change and the future of Europe. Springer. https://doi.org/10.1007/978-3-031-23328-9_7. Accessed 18 Oct 2024

Tomas A (2024) Geothermal power in Iceland. https://guidetoiceland.is/best-of-iceland/geothermal-power-in-iceland. Accessed 5 Dec 2024

Chapter 3
The EU-27: Leaders in Emission Trading and Key Sectors

Keywords EU · ETS · Green · Netherlands · Paris · Carbon · Coal · Nuclear · Growth · Energy

3.1 Emission Patterns and Trends

The European Union has exhibited significant reductions in ghg emissions for more than three decades. There were particularly large decreases in the early 2020s. These emission declines, moreover, occurred during a period of economic growth. *While its emissions decreased by 37%, EU-27 GDP increased in "real" inflation-adjusted terms by 68%* (European Commission 2024a, b, c, d, e). See chapter Annex 3.2 for details.

The emission decreases are documented in Table 3.1.

3.2 Policy Overview

European countries have long been recognized for being climate action leaders. For example, CCPI (2024) ranked the Netherlands 1st in the world, Finland 2nd, Denmark 4th, and Estonia 6th. As a group of 27, the EU was ranked 5th.

The continuing EU-27, after the UK's departure in January 2020, has strengthened existing policies and expanded the types and coverage of its climate actions. The policy expansions are evident in the Emission Trading System (ETS), the Green New Deal, the European Climate Law, and the Fit for 55 legislation.

T. L. Brewer, *Climate Action Leaders*, SpringerBriefs in Energy,
https://doi.org/10.1007/978-3-031-95667-6_3

Table 3.1 EU-27 emissions (1990–2023)

	1990	1995	2000	2005	2010	2015	2020	2022	2023
All ghg (kt $CO2_e$)	4.6	4.2	4.1	4.2	3.8	3.5	3.1	3.2	2.9
Percent change since 1990	0.0	−8.8	−10.9	−9.8	−17.8	−25.0	−34.4	−31	−37

Source Excerpts from European Environment Agency (2024a, b)

3.2.1 Emission Trading System (ETS)

The Emission Trading System (ETS) was established in 2005 as the first international ETS, and it has subsequently been expanded in several phases (European Commission 2021b, 2022). By the early 2020s, it covered nearly half of EU ghg emissions, and a new set of regulations approved in 2023 expanded the sectoral coverage to include maritime transportation, buildings, and road transportation in a separate new ETS. A new set of regulations concerning Carbon Border Adjustment Measures (CBAMs) began a "transitional phase" in October 2023 with a "definitive period" beginning in January 2026. The purpose is to "address the social and distributional impact" of the new ETS, with offsets to compensate for carbon-intensive imports that compete with EU goods in the ETS.

3.2.2 Other Initiatives and Legislation

A wide range of other initiatives have also been undertaken, many of them based on "targets" for 2030 and related strategies and actions to achieve the targets. The Green New Deal and the European Climate Law were developed in the 2019–2020 period, with the former establishing generalized principles and goals and the latter providing more specific legal obligations.

The Green New Deal of 2020 explicitly included the goal of being the first continent to be carbon neutral by 2050. It also stated more broadly that "Climate change and environmental degradation are an existential threat to Europe and the world. To overcome these challenges, the European Green Deal will transform the EU into a modern, resource-efficient, and competitive economy, ensuring: no net emissions of greenhouse gases by 2050; economic growth decoupled from resource use; no person and no place left behind" (European Commission 2021a).

The European Climate Law of 2019 established a legally binding target of net zero greenhouse gas emissions by 2050. For 2030, it set a goal of reducing net ghg emissions by 55% compared with 1990 (European Commission 2024b). The Climate Law also provides for "measures to keep track of progress and adjust … actions accordingly."

The EU Fit for 55 is a large package of legislation that refers to the 55% reduction in ghg emissions by 2030 that is established as a legal obligation by the European Climate Law. The Fit for 55 has been described as "[underpinning] the EU's position as leading the way in the global fight against climate change" (European Council 2021). It has also been described as "landmark steps, showcasing the EU's commitment to transforming its climate goals into tangible, enforceable actions and setting a new standard in international governance" (Anderson 2022). The legislative package was proposed by the European Commission in July 2021 and passed by the Council and Parliament in October 2023 (with a temporary exception concerning energy taxation).

In addition to provisions about changes in the Emissions Trading System, the Fit for 55 package includes a long list of specific actions to be taken in many economic sectors and industry sub-sectors. The following examples are from the energy and transportation sectors (compiled from European Commission 2023a, b, c; European Council 2021; Anderson 2022).

A new package proposed by the Commission in February 2025—the "Green Deal Industrial Plan"—was intended to integrate policies that were supportive of industries with climate change policy. See Annex 3.2 for details of the proposal.

Energy

Energy sector methane emission reductions. Following the 2020 EU Methane Strategy and COP26, the EU helped establish the Global Methane Pledge with a commitment to reduce methane emissions by 30% by 2030 from 2020. *The Fit for 55* provisions include reductions in venting and flaring at fossil fuel extraction and distribution facilities, greater methane leakage detection, and more stringent industry reporting of methane emissions.

Hydrogen and decarbonized gas market. Supports a transition away from fossil fuels by creating a competitive renewable gas market with secure supplies and consumer protections.

Transportation

Motor vehicles. 100% reductions of emissions by new cars and vans by 2035.

Shipping. A project named "FuelEU maritime initiative" has the objective of reducing the ghg emissions by 80% *by 2050*. (This can be done by switching out of diesel fuel into bio-fuels, electric propulsion systems, and wind-based technologies.)

3.3 Examples of National and Local Actions

The Treaty on the Functioning of the European Union provides for shared competence between the EU institutions and the individual member countries (European Commission 2024e). In that context, some national governments and local governments have undertaken climate actions on their own initiatives. In addition to the Nordic countries, who are either members or other participants in EU programs noted in the previous chapter, the Netherlands has been conspicuous among national governments and Paris among local governments.

3.3.1 Netherlands

The Netherlands has long been recognized as a climate action leader. This has been true, in part, because of the country's two basic geo-physical features: its terrain is relatively flat and part of it is below sea level. Its government's official succinctly stated perspective is:

A series of Dutch national agreements have created the core national framework for climate change policy (Netherlands 2025): the national Climate Act; the Climate Agreement; and the Climate Plans. There are two-year and five year-reviews—for the Climate Act and Climate Plan, respectively.

The Netherlands has also committed to several international agreements on tackling climate change, including the 1992 UN Framework Convention on Climate Change, the 1997 Kyoto Protocol, and the 2015 UN Climate Agreement at the COP in Paris. These formal public commitments reflect the view that "International cooperation is the best way to tackle greenhouse gas emissions and halt global warming" (Netherlands 2025).

3.3.2 Paris

Paris won a UN Global Climate Action Award in 2023 (UN Climate Change 2023a, b; also, C40 2025). A landmark in its formal policy-making process was the adoption in 2007 of a Climate Action Plan (C40 2018, 2020). The plan was revised in 2018 (European Commission 2020). At the time of the 2018 proposed revision, the city commissioned a study that explained the Action Plan, and an analysis of its expected effects. Although only 70,000 voted on the proposed revision, 95% voted in favor. The city has also developed a Climate Action Charter to facilitate partnerships with corporations and other organizations in the city (C40 2025.)

During the period 2004–2018, ghg emissions were reduced by 25%. Its "carbon footprint"—based on emission involved in consumption by the city, as opposed only to its production of emissions—was reduced 20% during the same period (UN

Climate Change 2023a, b; C40 2025.) When the Climate Action Plan was revised in 2018, it included actions to be taken during 2020–2030, toward a goal of reducing the city's emissions by 100% in 2050. The plan has been touted as having more than 500 actions (C40 2025). Its actions focus on the energy, transportation, and buildings sectors.

In the *energy* sector, more than 76,500 m^2 of rooftop solar panels have been installed, with an expectation that 20% of the city's rooftops would eventually be equipped with solar panels (UN Climate Change 2023a, b).

In the *transportation* sector, the city's plan is to phase out diesel-powered and gasoline powered vehicles by 2030.

In *housing*, by 2050 energy efficiency renovations to half the housing units are planned, with 65,000 already done by 2022 (UN Climate Change 2023a, b).

The third Paris Climate Plan (2018) defines an operational action plan for 2020–2030 in terms of the reduction of emissions and energy consumption, the development of renewable energy, adaptation to climate change and the implementation of carbon offsetting and sequestration tools to accelerate the local transition. By 2050, the city aims to reduce local emissions by 100%, achieving the goal of zero emissions in Paris, promote an 80% reduction in the carbon footprint of Paris compared to 2004 levels, and involve all local stakeholders in offsetting for residual emissions in order to achieve the zero net carbon target for the Paris area.

Black Carbon (BC) is another feature of Paris' climate change issues that has received attention from French government laboratories near Paris (Airparif 2021). Measurements from monitoring stations in and around the city during 2015–2019 found that that the highest levels of BC emissions were found near road traffic, including the Boulevard Periferique, Boulevard Haussmann, and Route A1. However, the overall level of local emissions during that period declined by 20–40%.

At the same time, there was also BC drifting into Paris from across the English Channel (i.e. La Manche), thus adding to local sources. Such incoming BC from a foreign source inevitably adds to the total air pollutants in the urban area, and is thus a health threat as well as a contributor to global warming. This combination thus creates *opportunities* to achieve the co-benefits of actions to reduce the emissions. It should also be noted that this phenomenon is not the same as the issues posed by the distinction between *domestically-based* ghg emissions resulting from *domestic production* versus *consumption based* emissions that include the emissions of imports—discussed below in Annex 3.1.

Annex 3.2 focuses on the proposal for an EU Green Deal Industrial Plan, and Annex 3.3 focusses on the process of the UK ending domestically-produced, coal-fired electric power plants, which included a period when the UK was a member of the EU.

3.4 Summary

The EU as a group of 27 countries has established a leading role globally in actions to address climate change problems and alternative solutions. In addition to its focus on an emission trading system as a central component of its array of actions, it has also addressed specific issues in the key sectors of energy and transportation. It has also begun to combine climate change policies with long-term economic restructuring and growth initiatives.

At the same time the EU-27 has been pursuing a range of alternative actions, national and local governments have been implementing action policies that are consistent with EU actions—with examples in the Netherlands' coastal and waterway projects and in Paris' transportation policies. The provisions for such shared competence between EU and its members is provided in the Treaty on the Functioning of the European Union.

Annex 3.1: Changes in Emissions and Changes in GDP

A key issue about emission reductions concerns their relationship with economic growth—in particular, whether the former reduces the latter. The answers depend on many complicating factors, such as short-term versus long-effects. In any case, an unavoidable complication is whether the change in emissions is based on "production" or "consumption" accounting, as discussed in Chap. 1. Table 3.2 provides evidence of the importance of the distinction between consumption-based and production-based accounting; it illustrates the different impressions about the relationships of emission reductions to GDP growth. Although the table does not provide a conclusive answer to the issues, it does illustrate the dependence of the answer on whether the indicator of the emission change is consumption-based and production-based accounting. For the EU-27 and the individual European countries in the table, there are substantial decreases in emissions, whether measured on a consumption-basis or production-basis, though the extent of the reductions obviously differ between the two indicators. However, the results for the U.S.—for a non-European comparison—are significantly different: *the consumption change is a large increase, while the production change is a very small decrease.* In short, the table reveals a compelling reminder of the consumption-production distinction.

Annex 3.2: The Green Deal Industrial Plan

The Commission President's summary of the EU's "Green Deal Industrial Plan" emphasized its place in the evolution of EU climate change policymaking: "We have a once in a generation an opportunity to show the way with speed, ambition and a sense of purpose to secure the EU's industrial lead in the fast-growing net-zero technology sector. Europe is determined to lead the clean tech revolution."

Annex 3.3: The Rise and Demise of Coal-Produced Electricity in the UK: 1880–2025

Over its nearly century-and-a-half history, the changing shares of production of electricity from coal were directly influenced by a combination of economic changes and technological changes, as well as changes in government policies. The policy changes occurred *at the regional (EU) level, as well as* UK national and local levels.

Sources of Electricity

By 1990, coal was the source of about two-thirds of UK electricity production, but a decade later it was only about one-third. Table 3.3 documents the demise of coal-fired electrical power plants in the early twenty-first century. The table also tracks oil and natural gas during that period. Although oil was low and declining, natural gas increased significantly, so that by 2020 it was the source of about as much as coal had been in 2010. Overall, the amount of electric power from fossil fuel plants declined from 145.5 TWh in 2000 to 103.1 in 2023—a decrease of 29.1% (Table 3.3).

Policies

The box below highlights key policy changes, including those of the EU that directly affected UK policies during its membership. Although the UK was only in the EU for about one-third of the 124-year history of the UK coal industry, EU policies played an important role on two occasions (1988 and 2005).

Table 3.2 Indicators of GDP compared with consumption-based and production-based emissions

Countries	Percentage change (1990–2022)			
	GDP	Accounting basis of emission changes		
		Consumption	Production	Emission decrease—Ratio: Cons/Prod (%)
EU-27	+70.94	−21.18	−29.20	72.5
France	+62.27	−17.22	−25.45	67.7
Germany	+56.68	−30.32	−36.34	83.4
Sweden	+92.65	−20.05	−34.85	57.5
UK	+82.37	−27.01	−47.86	56.4
US	+119.14	*+11.82*	*−00.82*	*1441.5*

Source Excerpted from IEA (2024), except ratios computed by author

Table 3.3 Sources of UK electric power production

Source	Electricity amounts generated (TWh)				
	2000	2010	2020	2023	Percent change 2000–2023 (%)
Coal	122.3	108.7	6.5	4.5[a]	−96.3[a]
Gas	14.8	175.7	111.6	96.2	+550.0
Oil	8.4	5.5	1.4	2.4	−42.8
Fossil fuels	145.5	289.9	119.5	103.1	−29.1

[a]It was 0.0 by 2025

Sources Excerpted and compiled by the author from IEA (2024); Lempiere and Evans (2024); Our World in Data (2024)

Box 3.1 Chronology of UK Policy Changes and Reports

1988—EU Large Combustion Plant Directive of 1988 required reductions of sulphur dioxide emissions, and which was expanded in 2008 with more restrictive limits on power plants' emissions.

2000—UK Royal Commission on Environmental Pollution—The government should limit ghg emissions in 2050 to 60% less than 2000, and it should engage in a "rapid deployment of alternative energy sources" as alternatives to fossil fuels.

2003—UK Department of Trade and Industry (DTI 2003) produced a report titled "Our energy future—creating a low carbon economy," which left open the possibility of new coal-fired plants if they included carbon capture and storage (CCS). By the time of the 2003 energy white paper, the "60% by 2050" target was government policy, as was a goal for 10% of electricity to be renewable by 2010, supported by a "renewables obligation." New nuclear was "not rule[d] out"—but it remained uncertain. Yet the 2003 white paper also left the door open to "cleaner coal" using carbon capture and storage (CCS). And it proposed government-backed investment in new coal reserves.

2005—Entry into force of the *EU's Emission Trading System*, which included a provision that power plants should pay a fee for carbon dioxide emissions.

2008—UK Climate Change Act passed. It included the first country's legally-binding climate change goal.

2012—Plan for a new coal-fired plant was cancelled, and no new coal-fired plants were subsequently built.

2013—Energy Act established new emission performance standards at 450 g CO_2 per kilowatt hour for new plants, which was about half of the emissions of an unabated coal plant. A carbon price floor was established and made gas plants more attractive than coal plants.

2020—UK government announced that the last coal-fired plant would close by the end of 2024.

2024—Last coal-fired plant closed on September 30.

Sources: IEA (2024); Lempriere and Evans (2024); Our World in Data (2024)

Changes in Coal Production and Trade

Another important aspect of changes in the UK's consumption of coal is the change from being a coal exporter to a coal importer. This is directly relevant to the levels of reported ghg emissions. While the emissions from domestically produced coal are included in standardized international reporting systems, emissions from imported coal are not because they are obviously produced in another country. They are therefore included in that *exporting* countries' emissions from their *production.* Thus, as the UK shifted from domestically produced coal to imported coal to fire its electric power plants, the officially reported emissions of that sector of the economy declined more rapidly than the decline in the coal-fired power stations.

The following report (UK 2024) illustrates the complexities and instabilities of these trends: "Coal demand for coal-fired electricity generation rose from 127 thousand tonnes in Quarter 2 2023 to 135 thousand tonnes in Quarter 2 2024." This demand, however, was being fulfilled by imported coal, not domestically produced coal.

Note also that a similar emerging contemporary issue is implicit in the same report—with italics added: "Coal use is being phased out with electricity generation favouring gas, nuclear and renewables. As coal use is being phased out, electricity generation favours gas, nuclear and renewables and, more recently, *imported electricity.*" This observation about the trend toward *imported electricity* raises an obvious question: How is it produced?

The climate change significance of these technical measurement and reporting systems is that the reported emissions—which are based on domestic *production* and do not reflect *consumption*—are sometimes misinterpreted. In the broader context of international comparisons of countries' emissions, comparisons between emissions from national production and emissions from national consumption pose significant international political and ethical issues, in addition to national economic accounting practices (Afionis et al. 2017; Chen et al. 2018; Grasso 2016; Liu 2015). A single example from the UK is that the percentage change in UK per capita emissions from *production* between 1990 and 2020 was minus 53.8% as compared with only minus 41.3% in *consumption.*

These issues are part of a larger set of issues about interactions in the relationships of policies concerning (a) international trade patterns and trends in goods and services with (b) government actions concerning climate change (Brewer 2008; World Bank 2024; World Economic Forum 2022). The issues include international direct investment and associated international technology flows.

References

Afionis S, Sakai M, Scott K, Barrett J, Gouldson A (2017) Consumption-based carbon accounting: does it have a future? Wiley Interdiscip Rev Clim Change 8:e438. Accessed 1 Nov 2024

Airparif (2021) Black carbon monitoring in Ile-De-France. https://www.airparif.fr/en/actualite/2021/la-surveillance-du-carbone-suie-en-ile-de-france. Accessed 26 Jan 2025

Anderson K (2022) Everything you need to know about the EU's 'Fit for 55' package. https://greenly.earth/en-us/blog/ecology-news/all-you-need-to-know-about-the-fit-for-55-plan. Accessed 19 Jan 2025

Brewer T (2008) The trade and climate change joint agenda. CEPS Working Document No. 295. http://www.ceps.eu. Accessed 1 Nov 2024

C40 (2018) Paris climate action plan: towards a carbon neutral city and 100% renewable energy. https://www.c40knowledgehub.org/s/article/Paris-Climate-Action-Plan-Towards-a-carbon-neutral-city-and-100-renewable-energy?language=en_US. Accessed 25 Jan 2025

C40 (2020) Climate action planning framework. https://www.c40knowledgehub.org/s/article/Climate-Action-Planning-Framework?language=en_US. Accessed 25 Jan 2025

C40 (2025) C40 knowledge hub. https://www.c40knowledgehub.org/s/?language=en_US. Accessed 25 Jan 2025

Chen Z-M et al (2018) Consumption-based greenhouse gas emissions accounting with capital stock change highlights dynamics of fast-developing countries. Nat Commun 9:3581. https://www.nature.com/articles/s41467-018-05905-y#Abs1. Accessed 1 Nov 2024

European Commission (2020) Revised thermal regulations [of France] (2020). https://commission.europa.eu/projects/revised-thermal-regulation-re2020_en. Accessed 4 July 2024

European Commission (2021a) The European green deal—delivering the EU's 2030 climate targets. https://commission.europa.eu/strategy-and-policy/priorities-2019-2024/european-green-deal/delivering-european-green-deal_en. Accessed 16 Jan 2025

European Commission (2021b) Supporting climate action through the EU budget. https://climate.ec.europa.eu/eu-action/eu-funding-climate-action/supporting-climate-action-through-eu-budget_en. Accessed 18 Jan 2025

European Commission (2022) Climate action: EU funding for climate action. https://commission.europa.eu/legal-notice_en. Accessed 18 Jan 2025

European Commission (2023a) Commission welcomes completion of key 'Fit for 55' legislation, putting EU on track to exceed 2030 targets. https://ec.europa.eu/commission/presscorner/api/files/document/print/en/ip_23_4754/IP_23_4754_EN.pdf. Accessed 21 Jan 2025

European Commission (2023b) Delivering the European green deal. https://commission.europa.eu/strategy-and-policy/priorities-2019-2024/european-green-deal/delivering-european-green-deal_en. Accessed 21 Jan 2025

European Commission (2023c) Legal documents on delivering the European green deal. https://commission.europa.eu/publications/legal-documents-delivering-european-green-deal_en. Accessed 21 Jan 2025

European Commission (2024a) EU climate action progress report 2024: report from the commission to the European parliament and the council. https://data.consilium.europa.eu/doc/document/ST-7445-2024-INIT/en/pdf. Accessed 16 Jan 2025

European Commission (2024b) Climate action. https://climate.ec.europa.eu/index_en. Accessed 16 Jan 2025

European Commission (2024c) Carbon border adjustment mechanism (CBAM) questions and answers. https://taxation-customs.ec.europa.eu/system/files/2023-12/Questions%20and%20Answers_Carbon%20Border%20Adjustment%20Mechanism%20%28CBAM%29.pdf?trk=public_post_comment-text. Accessed 4 June 2024

European Commission (2024d) Climate Action. https://climate.ec.europa.eu/index_en. Accessed January 16, 2025.

European Commission (2024e) EU competencies in the field of climate action. https://climate.ec.europa.eu/eu-action/eu-competences-field-climate-action_en. Accessed 16 Jan 2025

European Council (2021) Fit for 55. https://www.consilium.europa.eu/en/policies/fit-for-55/. Accessed 16 Jan 2025

European Environment Agency (EEA) (2024a) EEA greenhouse gases—data viewer. https://www.eea.europa.eu/data-and-maps/data/data-viewers/greenhouse-gases-viewer. Accessed 4 July 2024

European Environment Agency (EEA) (2024b) Total net greenhouse gas emission trends and projections in Europe. https://www.eea.europa.eu/en/analysis/publica. Accessed 16 Jan 2025

Grasso M (2016) The political feasibility of consumption-based carbon accounting. New Pol Econ 21:401–413. Accessed 1 Nov 2024

IEA (2024) United Kingdom, sources of electricity production. https://www.iea.org/countries/united-kingdom/electricity. Accessed 1 Nov 2024

Lempriere M, Evans S (2024) Q&A: how the UK became the first G7 country to phase out coal power. https://interactive.carbonbrief.org/coal-phaseout-UK/. Accessed 9 Oct 2024

Liu L (2015) A critical examination of the consumption-based accounting approach: has the blaming of consumers gone too far? Wiley Interdiscip Rev Clim Change 6:1–8. Accessed 1 Nov 2024

Netherlands (2025) Dutch vision on global climate action. https://www.government.nl/topics/climate-change/dutch-vision-on-global-climate-action Accessed 25 Jan 2024

Our World in Data (2024) Electricity generation from coal. https://ourworldindata.org/grapher/electricity-coal?tab=chart&country=GBR. Accessed 1 Nov 2024

Paris Climate Plan (2018) https://cdn.paris.fr/paris/2019/07/24/1a706797eac9982aec6b767c56449240.pdf. Accessed 15 July 2025.

UK Department for Energy Security & Net Zero (2024) Energy trends. https://assets.publishing.service.gov.uk/media/66f42056080bdf716392e8be/Energy_Trends_September_2024.pdf. Accessed 1 Nov 2024

UK Department of Trade and Industry (DTI) (2003) Our energy future—creating a low carbon economy. https://assets.publishing.service.gov.uk/media/5a7c1f5940f0b645ba3c6d4f/5761.pdf. Accessed 9 Oct 2024

UN Climate Change (2023a) Global climate action. https://unfccc.int/climate-action. Accessed 25 Jan 2024

UN Climate Change (2023b) City of Paris: carbon neutral by 2050 for a fair, inclusive and resilient transition. https://unfccc.int/climate-action/un-global-climate-action-awards/climate-leaders/city-of-paris. Accessed 25 Jan 2024

World Bank (2024) Trade and climate change. https://www.worldbank.org/en/topic/trade/brief/trade-and-climate-change. Accessed 1 Nov 2024

World Economic Forum (2022). https://www.weforum.org/stories/2022/02/rethinking-trades-relationship-to-the-fight-against-climate-change/. Accessed 1 Nov 2024

Chapter 4
Federal Political Systems: Brazil, Canada, Germany, Malaysia and Other ASEAN Countries

Keywords Brazil · Canada · Federalism · Provinces · States · Cities · Municipalities · Coasts · Deforestation · Territories

4.1 Introduction

Regional and local governments have been taking climate policy actions for many years, particularly in countries that have "federal" political systems. There are more than twenty countries with such systems, and they vary greatly in important aspects of their basic legal structure. The federalist systems of Brazil, Canada, Germany, and Malaysia are the focus of this chapter, and they illustrate many kinds of climate policy actions at all levels of governance as well as differences in their federalist structures.

The country-specific sections below highlight features of their federal systems and provide the national context and illustrative examples of their regional and local governmental climate policy actions. The case illustrations in these countries provide examples that could be adapted to political systems in other countries—non-federal as well as federal political systems. However, all four have long sea coasts, which affect their climate issues in terms of both mitigation and adaptation issues, as discussed in the chapter Annex.

4.2 Brazil

Brazil can be considered a rapidly emerging climate action leader—since January 2023 and the inauguration of a new government. For instance, during that year, the national government achieved a reduction of about one-third in the rate of Amazon

T. L. Brewer, *Climate Action Leaders*, SpringerBriefs in Energy,
https://doi.org/10.1007/978-3-031-95667-6_4

deforestation (Climate Action Tracker 2024). This and other developments at the national, regional, and local levels were receiving much international attention before the COP-30 meeting in Belem, Brazil, which is located in the Amazon rain forest.

4.2.1 National Policy Context

At the national government level, four sectors are central to Brazil's climate action agenda—namely, energy, transportation, agriculture, and of course forests. It has been noted that "Brazil is in an exceptional position to benefit from climate action, as it has several competitive advantages—most notably a relatively low-carbon energy supply, with large additional renewable energy potential" (World Bank 2023).

Energy

Brazil has developed an unusually low-carbon electric power generating industry, as Table 4.1 indicates. Its natural resources are evident in the proportions of hydro and bio-fuels.

Transportation

Brazil has supported bio-ethanol and bio-diesel fuels manufactured from its large sugar cane production. Nevertheless, the totality of the transportation sector is still emission intensive from traditional diesel-powered trucks and trains.

Agriculture

Both the livestock and cultivated segments of Brazil's agriculture sector are major producers of ghg emissions, including of course methane as well carbon dioxide. The CO_2e emssions of the sector were 490 megatonnes in 2020, and are not projected to decline much by 2030.

Table 4.1 Brazilian electric power sources (2020)

Source	Percent of total
Hydro	65.2
Wind	8.8
Solar	1.7
Nuclear	2.2
Fossil Fuels	13.0
Natural Gas	8.3
Coal	3.1
Oil	1.6

Source Compiled from World Bank (2023)

Forests

Because the deforestation of the Amazon and other Brazilian forests continues as a local, national, and global climate change problem, the need for the deforestation process to be reversed has been getting much attention at all governance levels in recent years. Although there has been a decline in deforestation in the past couple of years, there is a mixture of legal, economic, and political conditions that impede progress in further advances.

Another feature of Brazil's past and future climate policy agenda is the prominence of major cities as leaders in actions to reduce ghg emissions.

4.2.2 Municipal Policies

A report based on a survey of 5,570 Brazilian municipalities unsurprisingly concluded that it was important for them to develop varying mixtures in the types action that take into account the significant differences in their capacities to plan and enforce land use and other regulation (Brazil SEEG Municipios 2021).

Three state capital cities—Sao Paulo in the state of Sao Paulo, Belo Horizone in the state of Minas Gerais, and Porte Alegro in the state of Rio Grande do Sul—have become leaders in several sectors. Sao Paulo is the largest city in Brazil and South America; its metropolitan population is among the largest ten in the world. The three cities' climate change policies have been quite different, but all have been action leaders in Brazil (World Bank 2023).

Sao Paulo developed a climate action plan in 2019 based on an urban development planning and regulations plan from 2013. Belo Horizone regulations were put into effect in 2019-2020. Porte Alegro promulgated building regulations and incentives in 2020. All three cities have combined regulations that include land-use, zoning, mobility, building standards and modes of transportation (Table 4.2).

Table 4.2 Selected city and state populations

Capital city/state	Population (Millions 2022)		Location
	Capital city[a]	State (including capital)	
Sao Paulo/Sao Paulo	11.5 (municipal) 23.0 (metro)	41.3	Southeast coast
Belo Horizone/Minas Gerais	2.3 (municipal) 6.0 (metro)	19.6	South central
Porte Alegro/Rio Grande do Sul	1.4 (municipal) 4.2 (metro)	10.7	Southeast coast

Source World Bank (2023)

[a] The population of the *national* capital Rio de Jeneiro had a municipal population of 6.2 million and metropolitan population of 12.3 million (2022)

4.3 Canada

4.3.1 National Policy Context

In 2016, the Canadian national government's initial climate plan titled the "Pan-Canadian Framework on Clean Growth and Climate Change" included provinces, territories, and Indigenous peoples in the development process (Canada 2022, 2024a, b). Subsequent key national government decision-making processes in 2020 and 2022 were inclusive for the promulgation, respectively, of the "2030 Emissions Reduction Plan" and "A Healthy Environment and a Healthy Economy."

The latter in 2022 noted that "The Government recognizes that Indigenous climate leadership must be a cornerstone of Canada's climate actions and is partnering with First Nations, Inuit and Métis to set an agenda for climate action and a framework for collaboration." As for international collaboration, it noted that "Climate change is a global challenge that requires global solutions and Canada occupies a unique position in this challenge. ... Canada has a responsibility to take a leadership role through bold action to reduce greenhouse gas emissions...." In the midst of this series of national policy promulgations, with the support of four of the five party groups, the House of Commons declared a "national climate emergency," which required the country to achieve "clean growth and meaningfully reduce greenhouse gas emissions in all parts of the economy [which] are necessary to ensure a safer, healthier, cleaner, and more prosperous future for our children and grandchildren."

4.3.2 Provincial Climate Actions

Despite some legal differences, the regional governments in Canada typically include three "territories" in addition to the ten "provinces"—which are often included all-together in lists—as in the list below. Provincial populations are indicated in millions within the parentheses (Canada 2022, 2024a, b; British Columbia 2025; World Population Review 2025).

Alberta (4.8)

Carbon fee imposed on fossil fuels used in heating and transportation.

British Columbia (5.6)

In 2008, established the Innovative Clean Energy Fund to support clean energy technology projects, clean energy vehicles, research and development, and energy efficiency programs.

In 2022, began reductions of fossil fuels subsidies.

Supports purchases of zero emission vehicles.

Manitoba (1.5)

Biofuels program reduces transportation emissions.

New Brunswick (0.9)

Closed two fossil fuel power plants have been closed—one coal and one heavy oil.

Newfoundland and Labrador (0.5)

Expanding its hydroelectric power plants.

Nova Scotia (1.1)

In 2009 established limit on electricity ghg emissions (first in Canada).

Installed initial tidal turbine with connection to electrical grid.

Ontario (16.0)

In 2014, eliminated coal powered power plants (first in North America). In 2015, banned any future coal-fired plants.

Linked with Quebec and California cap and trade system in North America.

Prince Edward Island (0.2)

Largest wind energy penetration in Canada and 2nd highest in the world (after Denmark).

Biomass-fired district heating system since the 1980s.

Quebec (9.0)

Since 2014, in a cap-and-trade system with California and subsequently with Ontario.

Closed coal and oil power plants.

Northwest Territories (0.05)

Actions in development phase.

Nunavut (0.04)

In 2007, efficient energy systems installed in government-owned buildings.

Yukon (0.05)

Established Good Energy Residential Incentives Program, for purchases of high efficiency wood stoves, boilers and pellet stoves.

4.4 Germany

Because Germany is both a federal system and a member of the European Union, there are as many as four levels of government involved in some climate change policymaking processes. Such multi-level arrangements often include financial contributions from all levels. These possibilities have been significant in coastal protection programs, as indicated in the chapter Annex.

4.4.1 National Context

Germany was relatively early in its actions on climate change issues. Between 1990 and 2019, it reduced its ghg emissions 36 percent. By 2019, 43 percent of its electric power was generated by renewable sources—half of it by wind and most of the rest by solar. During a ten-year period, the national government actively supported 28,750 projects with expenditures close to 1 billion Euros (Germany 2024).

In 2019, the national parliament passed an ambitious "Climate Action Programme 2030," which included a wide range of plans and targets, especially in the energy, transportation, and buildings sectors. In 2023, the Renewable Energy Sources Act established a new legally binding target of 80 percent in the consumption of electricity by 2030.

An OECD (2022) review of actual policies, however, noted that in the transportation sector in particular, "Many opportunities such as broader use of speed limits, tolls for passenger and light duty vehicles, as well as congestion charges in urban areas have not been taken; others, such as, increased parking fees, are slowly materializing. The share of electric vehicles (EVs) is rapidly growing but remains modest in the total vehicle stock." It also noted that "Germany is one of the rare countries that does not tax vehicle purchase or registration."

However, since 1990, production-based greenhouse gas emissions have decreased in all German regions.

4.4.2 Regional and Local Actions

German regions decreased their emissions by 1.28% per year between 1990 and 2018. However, this is below the 1.93 percent annual rate required by the EU target of a 55 percent reduction by 2030 compared to 1990. There was a large difference between Schleswig–Holstein's low reduction rate of 8 percent and Saarland's high of 52 percent (OECD 2022).

See the chapter Annex for more information about the sub-national as well as national German government reactions to the problem of exposure to sea level rise and coastal flooding.

4.5 Malaysia and Other ASEAN Countries

"Shift for Sustainability: Climate Action Now" is the COP30 theme of Malaysia (2025). Its preliminary statement for the COP is that "Local communities play a critical role in protecting and managing natural resources" (Malaysian Green Technology Climate Change Centre 2024). The responsibilities for natural resources are noted in the constitution establishing a federal system—a responsibility that is especially important because of its long sea coast exposure at 4700 km. More generally, it is among the most vulnerable countries to the effects of climate change, not only because of its exposure to sea level rise and coastal flooding, but also because of its year round high temperatures and rainfalls as a tropical country.

At COP29, Malaysia announced a plan to reduce its carbon emissions 45 percent by 2030 compared with 2005. This is part of a broader Malaysian role as an international leader on climate change issues. The actions being taken are thus to mobilize international cooperation at both the regional and global levels. At the regional level, the Association of Southeast Asian Nations (ASEAN), including Malayia, has established eight "strategies" to address climate change, as summarized in Box 4.1.

Box 4.1 ASEAN's Strategies for Actions to Address Climate Change Issues

Accelerate green value chain integration. Establish cooperation frameworks and remove barriers to bring green products to market faster.

Support regional circular economy supply chains. Incorporate definitions and facilitate trade deals to support circular economy via regional trade.

Connect green infrastructure and markets. Build interconnectivity for green technologies and decarbonization solutions to underpin regional deployment.

Create interoperable carbon markets. Develop credible carbon markets that are interoperable with each other and global markets.

Create credible and common standards. Develop taxonomies and definitions on greenhouse gas reporting, carbon credits and energy efficiency that align with requirements of key markets.

Attract and deploy green capital. Build effective pathways to attract necessary capital to region and between member states.

Establish green best practice sharing. Share best practices in research-and-development, technology and green transition from within ASEAN and borrowing from the rest of the world.

It has been particularly active on international financial issues in both regional and global arenas. Regionally, Malaysia is Chair during 2025 for the Association of Southeast Asian Nations (ASEAN). Whereas Malaysia's population is 33 million, the total of all ten of the ASEAN countries is more than 600 million (Statista 2025)—and eight of them have seacoasts. Among the members, Indonesia's is the fourth longest coast in the world at 80,000 km.

4.6 Conclusion

These diverse federalist systems in different regions of the world illustrate the potential of sub-national local and regional governments to take actions that are effective in reducing emissions and that are economically, politically, and technologically feasible. The examples also illustrate how such actions can be adapted to highly variable local and regional circumstances. Yet, that is precisely an advantage they have in large countries with diverse regions.

Annex 4.1: Exposure to Sea Level Rise and Coastal Flooding

The localization of climate change issues in federalist government systems—and others—is especially pronounced in the case of sea level rise and coastal flooding. Although all affected areas share the obvious feature of their coastal location, there are many variations in other features, such as the land elevation above high tide, whether there is subsidence or rising of the coastal land mass, the distance of residential and other areas from the coast, the type and amount of economic activity in the area, and the sizes of the populations. Worldwide, 2.2 billion people—or 29 percent of the total world population as of 2018—lived within 50 km of a seacoast, which is 11 percent of the landmass area (Cosbey et al. 2024). Brazil's coast at 7500 km is the longest in South America. Canada's seacoast is the longest in the world at 202 thousand kilometers. Among the EU's 27 countries, 22 coastal countries have 68 thousand kilometers of coastline, of which Germany's is 2.4 thousand kilometers. Malasia's sea coast is 4700 km; of its 35 million population, several of its largest cities are coastal.

A localization feature of many of these coastal areas is that they emit large quantities of black carbon, which is the particulate matter that is among the three principal climate pollutants along with carbon dioxide and methane. On a per molecule basis, black carbon is thousands of times more potent than carbon dioxide as a global warming pollutant over a twenty-year period. Gasoline and diesel-powered motor vehicles are major sources of black carbon emissions. Because the atmospheric lifetime of black carbon is only about twelve days, reducing emissions of it are especially effective for reducing atmospheric warming.

Moreover, the black carbon particulates tend to fall to earth within a relatively small area. Though some molecules are also carried to much greater distances and thus act more like global warming agents such as carbon dioxide, *in the present context it is the localized concentration of black carbon particulates that is significant.*

Yet another feature is that black carbon emissions are also a major human health problem in the local area. Worldwide, black carbon emissions cause millions of deaths per year and many more cases of lung cancer and cardiovascular disease. In short, localized sea level rise and flooding as well as localized emissions of black carbon create incentives for localized actions. The actions reduce local property damage and save local lives.

In Canada, Prince Edward Island has partnered with the University of Prince Edward Island (UPEI) Climate Research Lab to study coastal vulnerability, including the award-winning Coastal Impacts Visualization Environment (CLIVE). CLIVE is an innovative 3D platform for visualizing the potential future impacts of *coastal erosion and coastal flooding* at local community scales, on PEI and elsewhere, using past data and Intergovernmental Panel on Climate Change models.

In Germany, a succinct summary for the European context of SLR follows from EEA (2025): "Most European coastal regions experience increases in both absolute sea level (as measured by satellites) and relative sea level (as measured by tide gauges), the latter being more relevant for coastal protection. There are sizeable differences in the rates of sea level change across Europe. Notably, sea levels relative to land along the northern Baltic Sea coast … are sinking. This is due to rising land levels caused by post-glacial rebound since the last ice age. … In future, relative sea level change along most of the European coastline is projected to be reasonably similar to the global average".

In Germany, in particular, there are coastal regions in both the North Sea and the Baltic Sea. According to the OECD (2022), "Sea-level rise and storms [in Germany] pose a high risk for the 3.2 million people who live in areas at risk of coastal flooding."

There are five coastal states (*Länder*), including the urban areas Hamburg and Bremen. The other three are mostly agricultural; one of them, Schleswig–Holstein, has been the subject of a case analysis (OECD 2022). A 2012 update of the Schleswig–Holstein Master Plan introduced a uniform flood safety standard of the 1-in-200-year event for the entire Schleswig–Holstein coast.

The sharing of legal responsibilities among three levels of government is evident in the German constitution, which defines coastal protection as a "joint task" for all citizens … with associated responsibilities, including financial, distributed among three levels of government (*federal* [*Bund*], *state* [*Land*] and *municipal* [*Gemeinde*]) by federal and state laws. In addition, the EU also has responsibilities. A brief tabulation of multi-level expenditures coastal protection in Schleswig–Holstein illustrates the extent of the distribution for both investments and maintenance of physical projects (OECD 2022). From 1962 [until 2024], spending on coastal protection in Schleswig–Holstein was EUR 2.73 billion. From the new Master Plan in 2001 [until 2024], the shares were about 50% from the state government, 37% from the national government, and 13% from the EU. Examples of projects include outer embankment slopes and a widening of dike crowns from 2.5 to 5.0 m during their reinforcement. State dikes protect approximately 90% of the total flood risk areas in Schleswig–Holstein up to the standard of 1-in-200-year event.

In Malaysia, sea level rise poses imminent dangers because of its peninsular configuration. Among the eight ASEAN countries with seacoasts, Indonesia's is the longest at 80,000 km (the fourth longest coast in the world); in Indonesia, the capital city Jakarta is being moved to a new capital at Nusantara more than a thousand kilometers away because of the combination of subsidence on land and sea level rise in Jakarta.

References

Brazil SEEG Municipios (2021) Press release (English language version). http://seeg.eco.br/press-release. Accessed 14 Feb 2025

British Columbia (2025) CleanBC. https://cleanbc.gov.bc.ca/climate-actions/. Accessed 20 Mar 2025

Canada, Government of (2022) Canada's climate actions for a healthy environment and a healthy economy. https://www.canada.ca/en/services/environment/weather/climatechange/climate-plan/climate-plan-overview/actions-healthy-environment-economy.html. Accessed 14 Feb 2025

Canada, Government of (2024a) Canada's climate plans and targets. file:///E:/DRAFTS%202025/Canada%E2%80%99s%20climate%20plans%20and%20targets%20-%20Canada.ca.mhtml. Accessed 14 Feb 2025

Canada, Government of (2024b) 2030 Emissions reduction plan: clean air, strong economy. https://www.canada.ca/en/services/environment/weather/climatechange/climate-plan/climate-plan-overview/emissions-reduction-2030.html Accessed 14 Feb 2025

Climate Action Tracker (2024) Brazil. https://climateactiontracker.org/countries/brazil/ Accessed 15 Mar 2025

Cosbey et al (2024) Accelerating growth of human coastal populations at the global and continent levels: 2000–2018. https://www.nature.com/articles/s41598-024-73287-x. Accessed 15 July 2025

EEA (2025) Global and European sea level rise. file:///F:/Global%20and%20European%20sea%20level%20rise%20_%20European%20Environment%20Agency's%20home%20page.mhtml. Accessed 16 Mar 2025

Germany (2024) What is the German government doing for the climate? https://www.bundesregierung.de/breg-en/issues/climate-action/government-climate-policy-1779414. Accessed 16 Mar 2025

Malaysian Green Technology Climate Change Centre (MGTC) (2024) Malaysia leads ASEAN in 2025. https://www.mgtc.gov.my/2024/11/malaysia-leads-asean-in-2025-committed-to-slashing-carbon-emissions-by-45/. Accessed 20 Mar 2025

Malaysia (2025) Malaysia Pavilion. https://malaysiapavilion-cop.com/. Accessed 20 Mar 2025

OECD (2022) Regions and cities at a glance 2022. Germany. file:///E:/CAL%20GERMANY%20BY%20OECD%20c0553d2d-en.pdf. Accessed 16 Mar 2025

Statista (2025) Total population of the ASEAN countries. https://www.statista.com/statistics/796222/total-population-of-the-asean-countries/#:~:text=Indonesia%20is%20not%20only%20a,number%20of%20inhabitants%20by%20far. Accessed 20 Mar 2025
World Bank (2023) Brazil. Country climate and development report. https://openknowledge.worldbank.org/entities/publication/a713713d-0b47-4eb3-a162-be9a383c341b. Accessed 17 Mar 2025
World Population Review (2025) Canadian provinces. https://worldpopulationreview.com/canadian-provinces. Accessed 12 Mar 2025

Chapter 5
The Future: Agendas for More Actions

Keywords Future · Fossil · Methane · Particulates · Carbon · 2030 · 2050 · China · India · US

5.1 Introduction

This chapter focuses on the agendas for actions as of early 2025. The short-term agenda is until 2030, which is frequently specified in official documents as well as research reports. The period until 2050 is commonly used for the longer term. The distinction is especially important as of 2025 because of the widespread view among climate scientists that much progress in reducing emissions must be made by 2030 in order to avoid the most catastrophic consequences of the increasingly rapid temperature increases.

A central question about the future of governments' climate actions is what the three biggest emitters will do. China and India are both on paths that include increasing emissions from fossil fuels, but at the same time they are on paths as major contributors to the expansion of key technological shifts in the energy and transportation sectors.

As of early 2025, the national government of the United States is on a fundamentally different path, as it focusses on increasing support of fossil fuels, reducing support of state and local emission reduction policies, and withdrawing from international emission reduction agreements. At the same time, however, many state and local governments continue to expand their emission reduction policies, even with national government legal challenges to them.

Section 5.2 of the chapter provides additional information about salient features of these three major emitters' policies heading into the future.

T. L. Brewer, *Climate Action Leaders*, SpringerBriefs in Energy,
https://doi.org/10.1007/978-3-031-95667-6_5

5.2 Projected Paths of the Three Big Emitters

The emissions of the three big national emitters in recent years have totalled slightly over half of the world total, as measured by CO_2e emissions. Of the 52% in 2023, 33% were Chinese, 13% American, and 6% Indian. Differences in *per capita* emissions among the three are especially notable. US per capita CO_2e emissions at 17.6 tonnes (in 1999) were *twice* China's at 8.6 tonnes and *seven* times India's at 2.5 tonnes. However, the long-term trends have been quite different, with the US declining while China and India were both increasing over the nearly three decades during 1992–2019 (Vigna and Friedrich 2023).

5.2.1 China

While China's coal consumption, especially for electricity production, is expected to continue at high levels for several more years, it is also on track to expand its wind, solar, and nuclear energy installations. "In 2023, China commissioned as much solar PV as the entire world did in 2022 while its wind additions also grew by 66% year-on-year. Over the past five years, China also added 11 GW of nuclear power, by far the largest of any country in the world" (IEA 2024a). Chinese investments in these industries equaled one-third of the world total. As for produced goods, its production and export of solar cells, lithium batteries and electric vehicles are increasing. Its exports of EVs to Europe are becoming more competitive compared with European-made and US-made alternatives.

5.2.2 India

India's consumption of fossil fuels has been an earmark of its energy sector; as of 2022, they accounted for three-fourths of its total energy supply—45.9% from coal, 24.0% from oil, and 5.1% from natural gas (IEA 2024c). However, there has been a "surge in Indian clean energy investment in recent years. Spending reached USD 68 billion in 2023, up by nearly 40% from the 2016–2020 average. Almost half of this was devoted to low-emissions power generation, which includes solar PV. … Clean energy investment is on track to double by 2030 under today's policy settings" (IEA 2024b).

5.2.3 *United States*

In the context of the "federal" political system in the United States, the climate action policies of many state and local governments have become increasingly important. As for local governments, major cities within these and other states, as well as small towns and other local governmental units, have also been active. One domestic consequence of these developments has been increasing tensions between the national government's policies, on the one hand, and the policies of some state and local governments, on the other hand. In fact, the tensions have prompted constitutionally significant legal cases. The fate of these cases in the judicial system, including the Supreme Court, and issues about the enforcement of court decisions, have become part of a more encompassing issue about the future of the federal system. The future of government climate action policies in the United States is therefore partly dependent on the resolution of these legal cases about the federal political system.

5.3 Sectors

Of course, there are significant variations among economic sectors in the types, sources, and impacts of emissions. At the same time, there are also some similarities in such attributes across sectors. An obvious one that has been well documented in many studies is the importance of emissions resulting from the use of fossil fuels in the energy, transportation, industry, and building sectors.

Fossil Fuels

Climate action agendas will continue to include fossil fuels emissions for many more years. They are among the most important and most problematic issues for several reasons. One is their economic significance and political significance in many of the big emitting countries, including China and India; another is their importance in many industries. In varying combinations, oil, gas, and coal are directly involved in producing electricity; powering motor vehicles, airplanes, ships, and off road equipment; producing cement, steel, and plastics; and in heating and lighting buildings. An extensive review of the climate change literature (Fletcher et al. 2024) summarized that "The only honest strategy for today is radical, immediate cuts in fossil fuel use." One way to facilitate such cuts is to improve the monitoring of methane emissions associated with fossil fuel extraction, processing, and distribution. The use of satellites to monitor methane emissions by the US National Aeronautics and Space Administration with the Union of Concerned Scientists (UCS 2024), launched such a new satellite-based system that was producing extensive methane emission data in 2024.

In addition to reducing fossil fuels' contributions to global warming, a transition out of fossil fuels can bring *co-benefits* in terms of human health. These benefits can be achieved because of the reduction of the particulate matter black carbon (BC) that emanates from fossil fuel uses. An analysis of the 2040 climate target impact assessment estimates that achieving the 90% target could reduce premature deaths due to air pollution from 466,000 per year in 2015 to 196,000 per year in 2040. That is a 57.9% annual reduction in deaths, or potentially more than 6 million over a decade (EU Commission 2024).

Short-Lived Climate Pollutants

The emissions of the two short-lived climate pollutants—methane and black carbon—are among the three most common climate change emissions globally along with CO_2 of course (IEA 2024e).

Methane and black carbon are both short-lived climate pollutants (SLCPs)—also sometimes referred to as "short-lived climate forcers"—with relatively short atmospheric lifetimes compared to carbon dioxide: only about 7–12 days for black carbon and about 12 years for methane. In addition, their relatively large atmospheric warming impact (Global Warming Potential or GWP) per molecule is quite high compared with CO_2—about 3200 for black carbon at 20 years and about 83 for methane. It should also be noted that there are several alternative indicators of the potencies of black carbon and methane relative to carbon dioxide, but they all reflect the basic fact that both black carbon and methane are much more impactful than carbon dioxie per molecule.

Because SLCPs have such a strong warming effect compared with carbon dioxide per molecule in the atmosphere, reducing them can have relatively rapid effects on atmospheric temperatures (c2es 2024). The basic distinction between methane and black carbon, on the one hand, and carbon dioxide and other long-lived gasses, on the other hand, has been known for many years. Until recently, however, neither methane nor black carbon had been receiving as much attention by policymakers. Now, because both methane and black carbon have relatively short atmospheric lifetimes, the potential reductions of the two short-lived emissions have become especially appealing because they are responsive to the increasing recognition that current and projected reductions of long-lived carbon dioxide and nitrogen oxide will not be sufficient to avoid the most catastrophic scenarios.

Methane

The IEA (2024a, b, c, d) has succinctly summarized the significance of methane: "[M]ethane is a very potent greenhouse gas, and rapid and sustained cuts in emissions from the energy sector are imperative to limit global warming to 1.5 °C. Methane also affects air quality because it can lead to ground level (tropospheric) ozone, a dangerous pollutant. Methane leaks can also pose explosion and health hazards."

Another problem is that methane is being released into the atmosphere from under the Arctic's melting tundra surface. Because Arctic temperatures have been increasing about *four times* the global average for at least two decades, this has made that extensive area of the Arctic a net carbon source in recent years, whereas it used to be a carbon sink. Moreover, there are feedback effects such that the warming of the regional air temperatures causes additional tundra melt. In short, if the Arctic tundra continues to be a net carbon source because of its methane emissions rather than a sink, it will be a major shock to future trends in global temperatures.

Black Carbon

The extreme rate of Arctic temperature increases is partly a consequence of direct deposits of black carbon (BC) on the area's ice sheet surfaces. BC that lands on the surface of the ice sheets is a direct cause of the Arctic ice melt that directly adds water to the Atlantic Ocean and increases the rate of sea level rise in other oceans as well.

Sea level rise (SLR) is already happening around the world at an increasing rate because of rising ocean surface temperatures; however, the Arctic and Antarctic contributions are making further contributions. The SLR contributions to the North Atlantic are greater than those to the North Pacific; this is *one* of the reasons why the U.S. East Coast (i.e. Atlantic Coast) has a projected SLR of 25–36 cm (10–14 inches) by 2030 compared with 2000. The comparable increase on the U.S. West Coast (i.e. Pacific Coast) is 10–20 cm (4–8 inches).

Energy

An extensive WRI report on the "energy transition" by Layke (2025) notes that "With prices falling and technology more widely available, clean energy uptake is growing rapidly. Investment in new clean power now outpaces investment in fossil fuels. This represents tremendous progress—but affordable technology is only one piece of the energy transition puzzle." There are, indeed, "six key opportunities that leaders should embrace to unlock a global energy transition at the pace and scale the world needs." See Box 5.1.

Box 5.1 Ways to Facilitate the Energy Transition

Modernizing Power Systems

Electricity grids need extensive upgrades to be able to transmit clean energy from where it's produced (often in remote rural areas) to cities and manufacturing hubs. They need to be more efficient. And they need to withstand increasingly extreme weather.

Attracting, Shifting and Mobilizing Finance

In 2024, global investments in the energy transition surpassed $2 trillion for the first time, with electric transport, clean energy and grid infrastructure making up the largest share. [Nevertheless] finance for clean energy needs to increase drastically.

Securing Critical Minerals Responsibly

[T]he world needs to secure an abundant supply of critical minerals [such as copper, nickel, lithium and cobalt] in a way that's safe, respects human rights, and minimizes risks to people and the environment.

Shifting Demand from Large Energy Users

More collaboration within cities and between companies in all sectors—from retailers to industry and buildings—can shift demand and build a market for clean, abundant and reliable power.

Overcome Siting Hurdles

To start, communities' concerns need to be heard and addressed early on. … Also important is making sure that the benefits of clean energy development are indeed felt locally.

Cultivating the Energy Workforce of the Future

The world needs a new generation of manufacturers, energy workers and entrepreneurs equipped to build and operate 21st-century systems.

Source: Excerpts from Layke (2025)

Nuclear Energy

The future of nuclear energy is on the agenda of many European countries, including France, which has had the world's biggest nuclear power industry in terms of both the number of reactors and their electrical output. Finland, Sweden, and the UK also have civil nuclear power industries, though on a smaller scale.

5.4 Concluding Comments

The projections of the future of climate change and its consequences are daunting. The results of additional climate science research projects tend to project increasingly severe future scenarios. Yet, at the same time, there is some progress in the adoption of emission-reducing policies and technologies. There are in fact many paths that could significant reduce the severity of the most calamitous scenarios.

However, there is also a clear consensus of conclusions in the on-going climate science results that more action is needed and that it is needed now. More governments at all levels must change their paths to 'more-and-sooner."

The numerous and diverse tables and lists in the following Annex 5.1 provide many examples of both the problems and the solutions. Annex 5.2 provides lists of many resources for information to keep up with new developments about both problems and solutions.

Annex 5.1: Past Paths and Projected Paths into the Future

Emission Trends and Patterns

See Tables 5.1, 5.2 and 5.3.

"Nations must collectively commit to cutting 42% off annual greenhouse gas emissions by 2030 and 57% by 2035 in the next round of Nationally Determined Contributions (NDCs)—and back this up with rapid action—or the Paris Agreement's 1.5 °C goal will be gone within a few years…." UNEP (2024). (Table 5.4).

Climate Change Tipping Points

As many as 21 climate tipping points have been identified. They can be listed in several ways, including the types of events they involve and the threshold temperatures that could trigger them. Because the temperatures at which they are precipitated are important, there is much interest in establishing such temperatures, as in Table 5.5.

Table 5.1 CO_2 emissions per capita by major emitting countries, 1950–2020 (tonnes)[a]

Countries	1950	2000	2010	2020
World	2.4	4.1	4.8	4.5
China	0.1	2.9	6.4	7.6
India	0.2	0.9	1.3	1.7
United States	16.5	21.4	18.3	13.9

[a]Carbondioxide(CO_2) emissions from fossil fuels and industry
Land-use change is not included
Source Excerpted from Our World in Data (2025a)

Table 5.2 Ratio of major emitting countries' emissions per capita to world total

Countries	1950	2000	2010	2020
World	2.4	4.1	4.8	4.5
China	0.04	0.71	1.33	1.69
India	0.08	0.22	0.27	0.38
United States	6.88		3.81	3.09

Source Computed by the author from Table 5.1

Table 5.3 Projections of global emissions gaps (Gt CO_2e)[a]

Year	Projected Gt CO_2e	1.5° Gap Gt $C0_2e$	2.0° Gap Gt $C0_2e$
2030	57	24	16
2035	57	32	21

Source Excerpted from UNEP (2024)

Table 5.4 *Climate change tipping points*

Potential individual tipping point events	Threshold temp. Est
Ice melt	
Greenland ice sheet collapse	1.5
West Antarctic ice sheet collapse	1.5
Barents sea ice abrupt loss	1.6
Arctic summer sea ice loss	2.0
Mountain glaciers loss	2.0
Arctic winter sea ice collapse	6.3
East Antarctic ice sheet collapse	7.5
East Antarctic subglacial basins collapse	3.0
Other sea and oceanic events	
Atlantic meridional ocean circulation collapse	4.0
Labrador sea convection collapse	1.8
Boreal permafrost	
Abrupt thaw	1.5
Collapse	4.0
Boreal forest	
Gradual thaw	1.5
Southern dieback	4.0
Northern expansion	4.0

There are others that are not in the table because they are considered either "uncertain" or "unlikely." The ice melts in the Arctic and Antarctic regions are especially concerning because of their temporal proximity and their effects on global sea level rise (Igini 2024).

Effects of Climate Change on Economies and Human Health

Table 5.5 Renewable energy increases (TwH)

Type	2010	2020	2023	Increase (TwH) 2010–2023	Increase (Ratio: 2023/2010)
Hydro	2646.64	3430.11	4240.01	1,593.37	1.60
Wind	31.41	346.41	2325.31	2293.90	74.03
Solar	1.062	33.90	1641.58	1,640.51	1,545.74
Other	184.88	379.44	781.50	596.62	4.23
Total	2863.99	4189.86	8988.39	6,125.39	3.14

Source Excerpted from Our World in Data (2025b). Increases computed by the author

Box 5.2 Economics and Health Effects

Climate change is the biggest health threat facing humanity, the World Health Organization says, estimating that it will cause around a quarter of a million additional deaths each year in 2030–50. These will mainly be from malnutrition, malaria, diarrhoea and heat stress.

Climate change is raising the cost of living

"Companies face up to $120 billion in costs from environmental risks in their supply chains by 2026," according to research published in 2021 by CDP, a nonprofit that runs the world's largest environmental disclosure system. This will include increased costs for raw materials. Sea level rises could pose the biggest threat to global supply chains, potentially putting ports and coastal infrastructure out of action. The sea is home to most of our biodiversity, and 3 billion people globally rely on it for their livelihoods, according to the UN. However, carbon emissions from human activity are causing ocean warming, acidification and oxygen loss, putting large numbers of marine-related jobs at risk, it says.

Source: Excerpts from World Economic Forum (2022)

Box 5.3 Effects of Climate Change on Economies

The world stands to lose close to 10% of total economic value by mid-century if climate change stays on the currently-anticipated trajectory, and the Paris Agreement and 2050 net-zero emissions targets are not met.

- The largest impact of climate change is that it could wipe off up to 18% of GDP off the worldwide economy by 2050 if global temperatures rise by 3.2 °C, the Swiss Re Institute warns.
- Forecast of 10% based on temperature increases staying on the current trajectory and the Paris Agreement and net-zero emissions targets not being met.
- Figure could rise to 18% of GDP by mid-century if temperatures increase by 3.2 °C in the most severe scenario.

Recent scientific research indicates that current likely temperature-rise trajectories, supported by implementation of mitigation pledges, would entail 2.0–2.6 °C global warming by mid-century. We use this as the baseline to simulate the impact of rising temperatures over time, while also modelling for the uncertainties around most severe possible physical outcomes. The result is that global GDP would be 11–14% less than in a world without climate change (ie, 0 °C change). Under the same no climate change comparative, the Paris target too result in negative GDP impact, but less much so (–4.2%). We also consider a severe scenario in which temperatures rise by 3.2 °C by mid-century, with society doing nothing to combat climate change. In this scenario, the global economy would be 18% smaller than in a world without warming, reinforcing the imperative of, if anything, more action on climate change.

Source: Excerpt from Swiss re (2022)

Solutions

Box 5.4 Reducing Black Carbon: List of Possibilities from the World Economic Forum

Potential Next steps for BC-emitting countries:

Residential cooking,
Residential heating,
New on-road diesel vehicles,
New off-road diesel engines,
New in-use diesel vehicles and engines,
Brick kilns,
Coke ovens,
Natural gas flaring,
Agricultural burning,
Forestry burning.

Potential next steps for the global community:

Give each climate pollutant its due.
Identify, demonstrate and share best BC mitigation practices.
Foster South-South regional cooperation on common source types.
Source: Adapted from World Economic Forum (2024).

See Table 5.6.

Of the 24 cells where there are comparisons, there are only two aberrations (Finland in 2010 and the US in 2000).

Annex 5.2: Sources of Information for Following Future Developments

Tipping Points

Tipping Points can be listed in several ways, including the types of events they involve and the threshold temperatures that could trigger them.

Table 5.6 Decoupling of emission reductions and GDP growth

Countries	Percent change since 1990			
	2000	2010	2020	2023
Finland				
GDP	+21.03	+39.20	+42.82	+46.69
Emissions	−3.50	*+4.69*	−40.27	−50.44
France				
GDP	+20.58	+27.10	+27.56	+39.24
Emissions	−2.06	−8.89	−38.40	−40.60
Germany				
GDP	+17.95	+29.48	+42.41	+49.04
Emissions	−16.95	−22.22	−41.41	−46.72
Sweden				
GDP	+19.08	+38.11	+48.17	+ 56.00
Emissions	−7.88	−15.72	−47.26	−48.45
US				
GDP	+24.02	+34.75	+51.77	+68.05
Emissions	+ *5.64*[a]	−9.66	−31.28	−29.25
UK				
GDP	+26.06	+37.03	+36.73	+53.26
Emissions	−8.17	−22.57	−53.83	−57.66

Source Excerpted from Our World in Data (2025a)

Emissions Data

EU Copernicus—https://www.copernicus.eu
European Environment Agency (EAA)—https://www.eea.europa.eu
European Space Agency—https://www.esa.int
IEA—https://www.iea.org/topics/climate-change
IPCC—https://www.ipcc.ch
OECD—https://www.oecd.org/en/topics/climate change.html
Our World in Data—https://ourworldindata.org

Temperatures

EU Copernicus—https://www.copernicus.eu
IPCC—https://www.ipcc.ch
World Metrological Organisation (WMO)—https://wmo.int

International Organizations, Agreements and Negotiations

Institute for Sustainable Development and International Relations—https://www.iddri.org/en or Institut du Développement Durable et des Relations Internationales (IDDRI)—https://www.iddri.org/fr

Intergovernmental Panel on Climate Change (IPCC)—https://www.ipcc.ch/
International Institute for Sustainable Development (IISD)—https://www.iisd.org/
OECD—https://www.oecd.org/en/topics/climate-change.html
UN Environment Programme (UNEP)—https://www.unep.org/
UN Framework Convention on Climate Change (UNFCCC) https://unfccc.int/
UN Climate Change—https://www.un.org/en/climatechange/what-is-climate-change
World Bank—https://climateknowledgeportal.worldbank.org/

Governmental Policies

Climate Change Performance Index (CCPI) (2024)—https://ccpi.org
Climate Action Tracker—https://climateactiontracker.org/
OECD—https://www.oecd.org/en/publications/the-climate-actions-and-policies-measurement-framework_2caa60ce-en.html
European Commission—https://commission.europa.eu/index_en
IEA—https://climateknowledgeportal.worldbank.org/
OECD—https://www.oecd.org/en/topics/climate-change.html
World Bank—https://climateknowledgeportal.worldbank.org/
Center for Climate and Energy Solutions (c2es)—https://www.c2es.org/about/
Carbon Tracker— https://carbontracker.org/

NGOs

Center for Climate and Energy Solutions (c2es)—https://www.c2es.org
Center for European Policy Studies (CEPS)—http://www.ceps.eu
Climate Action—https://www.climateaction.org
Climate Change AI—https://www.climatechange.ai
C40 Knowledge Hub—https://www.c40knowledgehub.org/
Earth.org—https://earth.org
Global Goals—https://www.globalgoals.org
International Institute for Sustainable Development (IISD)—https://www.iisd.org/
Local Governments for Sustainability (ICLEI)—https://iclei.org/
Natural Resources Defense Fund (NRDC)—https://www.nrdc.org
Resources for the Future (RFF)—https://www.rff.org/
Union of Concerned Scientists (UCS)—https://www.ucs.org/
World Economic Forum—https://www3.weforum.org
World Resources Institute (WRI)—https://www.wri.org

Daily and weekly news sources

Associated Press News (AP)—https://apnews.com/
British Broadcasting Corporation (BBC)—https://www.bbc.com/news
Bloomberg—https://www.bloomberg.com/green
Economist—https://www.economist.com/
Financial Times (FT)—https://www.ft.com/

Georgetown University Climate Center (GCC)—https://www.georgetownclimate.org/
Inside Climate News—https://insideclimatenews.org/newsletter/
Le Monde (English)—https://www.lemonde.fr/en/
New York Times—https://www.nytimes.com/
Politico (EU)—https://www.politico.eu/
Politico (US)—https://www.politico.com/
Reuters—https://www.reuters.com
Washington Post—https://www.washingtonpost.com/

References

C2es (2024) Short-lived climate pollutants. https://www.c2es.org/content/short-lived-climate-pollutants/. Accessed 14 Nov 2024

EU Commission (2024) Report from the commission to the european parliament and the council. Climate Action Progress Report 2024. file:///e:/climate%20action%20leaders/leaders%20europe/com_2024_498_f1_report_from_commission_en_v4_p1_3729454.pdf. Accessed 3 Dec 2024

Fletcher et al (2024) Earth at risk: an urgent call to end the age of destruction and forge a just and sustainable future. PNAS Nexus. https://doi.org/10.1093/pnasnexus. Accessed 30 Dec 2024

ICLEI (2025) COP Event Guidelines. https://iclei.org/. Accessed 15 July 2025.

IEA (2024a) World energy investment 2024, China. https://www.iea.org/reports/world-energy-investment-2024. Accessed 5 Mar 2025

IEA (2024b) World Energy Investment 2024, India. https://www.iea.org/reports/world-energy-investment-2024. Accessed 5 Mar 2025

IEA (2024c) Coal 2024. https://www.iea.org/countries/india. Accessed 5 Mar 2025

IEA (2024d) Understanding methane emissions. https://www.iea.org/reports/global-methane-tracker-2024/understanding-methane-emissions. Accessed 5 Mar 2025

Igini M (2024) The tipping points of climate change: how will our world change? Earth.org. https://earth.org/tipping-points-of-climate-change/. Accessed 13 Mar 2025

Layke J (2025) 6 Opportunities to accelerate the global energy transition. https://www.wri.org. Accessed 15 Mar 2025

Our World in Data (2025a) Per capita CO_2 emissions. https://ourworldindata.org/explorers/co2?time=1750..latest&country=CHN~USA~IND~OWID_WRL&Gas+or+Warming=CO%E2%82%82&Accounting=Territorial&Fuel+or+Land+Use+Change=All+fossil+emissions&Count=Per+capita. Accessed 13 Mar 2025

Our World in Data (2025b) Renewable energy generation. https://ourworldindata.org/renewable-energy. Accessed 13 Mar 2025

Swiss re (2022) The economics of climate change: no action not an option. https://www.weforum.org/stories/2021/06/impact-climate-change-global-gdp/. Accessed 13 Mar 2025

UN Climate Change (2024a) Introduction to climate finance. https://unfccc.int/topics/introduction-to-climate-finance. Accessed 13 Nov 2024

UNEP (2024) Emissions gap report 2024. https://www.unep.org/resources/emissions-gap-report-2024;file:///F:/UNEP%202024%20EGR2024.pdf. Accessed 13 Mar 2025

UNFCCC (2024a) UN climate change conference Baku. https://unfccc.int/cop29. Accessed 28 Nov 2024

Vigna L, Friedrich J (2023) WRI. 9 charts explain per capita greenhouse gas emissions by country. https://www.wri.org/insights/charts-explain-per-capita-greenhouse-gas-emissions. Accessed 5 Mar 2025

World Economic Forum (2022) Climate action. Climate change is already altering everything, from fertility choices to insuring our homes. https://www.weforum.org/stories/2022/06/climate-change-weather-extreme-health/. Accessed 13 Mar 2025

World Economic Forum (2024) Black carbon reduction: a rapid action plan. https://www3.weforum.org/docs/WEF_Black_Carbon_Reduction_2024.pdf. Accessed 13 Mar 2025

Made in the USA
Monee, IL
14 September 2026

62974049R00046